P9-DTN-020

RENEWALS 691-4574

DATE DUE

WITHDRAWN
UTSA LIBRARIES

PROBABILITY METHODS IN OIL EXPLORATION

"KOX No. 1," a Lansing discovery well in northwestern Kansas, silhouetted against the evening sun. The well was drilled on a high probability anomaly which was verified by a seismic survey prior to drilling.

Probability Methods in Oil Exploration

John W. Harbaugh
Stanford University

John H. Doveton
Kansas Geological Survey

John C. Davis
Kansas Geological Survey

A Wiley-Interscience Publication

JOHN WILEY & SONS

New York · London · Sydney · Toronto

To Josephine, Mary, and Carolyn

Copyright © 1977 by John Wiley & Sons, Inc.

All rights reserved. Published simultaneously in Canada.

No part of this book may be reproduced by any means, nor transmitted, nor translated into a machine language without the written permission of the publisher.

Library of Congress Cataloging in Publication Data

Harbaugh, John Warvelle, 1926–
  Probability methods in oil exploration.

  "A Wiley-Interscience publication."
  Includes index.
  1. Prospecting—Statistical methods. 2. Petroleum.
3. Probabilities. I. Doveton, John H., 1944–
joint author. II. Davis, John C., 1938–   joint author.
III. Title.

TN271.P4H28   622'.18'282   76-50631
ISBN 0-471-35129-6

Printed in the United States of America

10 9 8 7 6 5 4 3 2 1

LIBRARY
University of Texas
At San Antonio

Preface

Prospecting for oil is a high-risk business because it involves strong elements of uncertainty and the odds of success are relatively marginal. The keystone of conventional exploration philosophy is that geologic information is the most effective guide to the discovery of oil, although geologic properties are inherently statistical. Their statistical nature is in part a consequence of the incomplete knowledge available through the limited number of wells and geophysical coverage, which causes uncertainty in prediction of subsurface variations between control points. In addition, the functional relationships between geology and the location of oil are complex and highly variable. In spite of the daunting problems caused by incomplete data and the complexities of interpretation, it is generally agreed that consideration of the relationships between subsurface properties and oil accumulation will continue to result in exploration strategies that are far more successful than alternatives not based on these considerations.

This book is principally concerned with methods for treating exploration data on a statistical or probabilistic basis. Our approach implies that there are cause-and-effect relationships between geologic conditions and the occurrence of oil, and is in keeping with the general exploration philosophy of the petroleum industry. The main purpose of our book, however, is to suggest methods by which the uncertain relationships between geology and oil occurrence can be treated from a formal statistical or probabilistic viewpoint.

The distinction between *uncertainty* and *probability* is a critical one. Most explorationists will concede that oil exploration is a highly uncertain business. Few explorationists, however, deal with uncertainty in a probabilistic sense. Our use of the term probability implies some knowledge of the statistical behavior of a variable; in contrast, uncertainty implies lack of knowledge about its statistical behavior. For example, if a geologic property is a random variable with a normal frequency distribution, its individual behavior can be predicted only on a statistical basis. However, its mean and variance provide an exact statistical description of its long-term or overall behavior. Thus, we remain uncertain if we have no knowledge of the statistical parameters of a variable, but we can treat the variable probabilistically if we can measure its statistical characteristics. Our thesis is that statistical parameters of geologic features that influence the occurrence of oil can be estimated, and that a formal probabilistic approach is both feasible and preferable as a strategy for oil exploration.

Typically, an explorationist evaluates evidence on the basis of industry experience, coupled with geologic theories of oil migration and accumulation. His qualitative appreciation is used to selectively weigh the data before him and to form a judgment on the strength of the evidence. If the evidence seems compelling, it will be reflected in his exploration decisions; the geologist will recommend or reject a play accordingly. This mental process is an unconscious, informal statistical analysis. Experience provides the probabilities, the relationship between geologic variables and oil occurrence, and a feel for the uncertainty caused by the limited amount of data at hand. In our research, we have attempted to parallel this pattern of analysis performed by geologists, but in a way that will result in *explicit* assessments of geologic prospects instead of intuitive, qualitative judgments. We have substituted an historical analysis of drilling success for experience, measured the statistical correlations between geologic variables and oil discovery, and analyzed the uncertainty which results from the spacing and pattern of control points. Because each step in the procedure is quantitative, the odds of success can be calculated explicitly, and relatively precise evaluations of different prospects can be formed. Although the procedures described in this book are mathematical and involve the use of computers, they have the sole purpose of emulating the decision process of human explorationists. However, by using quantitative techniques we obtain consistency, objectivity, and precision which should make exploration more efficient and rewarding.

Neither our methods, nor quantitative methods drawn from any other discipline, can change the fundamentally probabilistic nature of oil exploration. However, it does appear that the methods of mathematical probability can bring about an improvement in the efficiency of the exploration process. Even a proportionally small improvement in the efficiency of exploration will be highly important from an economic standpoint, both for an individual company as well as for the industry as a whole.

Previously, probability methods have been confined to what might more properly be called "basin evaluation" and to petroleum exploration studies that considered only a few geologic factors, such as the areas and volumes of oil fields in specific regions. These frequency distributions can be used to assess the probabilities attached to discovery of oil fields of varying magnitudes, but only on a regional basis, such as over a large geographic area or over an entire sedimentary basin. This information is only marginally useful in estimating the probabilities attached to a specific play or prospect, where decisions are necessarily conditioned by the particular geologic features at the prospect site and from which the play must be evaluated. Other probabilistic techniques involve the intuitive or subjective estimation of probabilities attached to estimates of oil occurrence on a regional basis. While also useful in some respects, intuitive estimates are fraught with uncertainty and may lend a false sense of security because the estimates are represented by numbers which in turn are subjected to involved mathematical manipulations. Still other probabilistic techniques have concentrated on the design of search strategies to locate targets distributed through space according to assumed spatial distributions. Although useful in some contexts, we think that such an approach is less powerful than one which uses the conditional relationship between geologic variables and the occurrence of oil. Thus, in this book we are principally concerned with methods for estimating probabilities attached to specific oil plays and prospects, utilizing pertinent geological and geophysical information to refine the evaluation.

Many of the examples in this book have been taken from studies that were part of the Kansas Oil Exploration (KOX) Project of the Kansas Geological Survey. The objective of the KOX Project has been to develop a computer-based system for guiding oil exploration programs in the Midcontinent region. More specifically, KOX has been designed to assist independent oil operators who do the bulk of the exploration in Kansas and elsewhere in the Midwest. Research has centered around the analysis of both structural and stratigraphic data, so the probability of success attached to exploration activities such as the drilling of a wildcat well can be estimated. The estimates are conditional on the subsurface geology as interpreted from the available data.

This book is directed to petroleum geologists and exploration managers who are concerned with appraising oil prospects and making exploration decisions. Because of their scale, the examples given are particularly pertinent to independents and drilling-fund operators, but the methodology is suitable for exploration activities that are national or international in scope. In addition to this audience of practicing explorationists, those interested in the more applied aspects of geomathematics, either as students or researchers, may find methods and philosophy of interest.

Several assumptions about the reader's background are incorporated in the book. First, we assume that the reader has at least an elementary knowledge of the principles and practice of petroleum geology and exploration. Second, we assume that the reader has a general appreciation of the elementary concepts of probability and statistics, but an advanced knowledge of mathematical probability is not assumed. Terms that may be unfamiliar are explained in a glossary. Our goal throughout has been to discuss the principles and practice of probability theory as applied to oil exploration, so these methods could be evaluated and implemented by the widest possible group of explorationists.

We acknowledge the help of many persons in preparing this book. Since much of the background stems from the KOX Project, we owe special thanks to those who participated in and who helped direct that project. Most of the computer programs, including the SURFACE II mapping system, were written by Mr. Robert Sampson of the Kansas Geological Survey. Mr. Harold Cable, with Mrs. Effat Peyrovian, was responsible for operating programs at the Kansas Survey. Dr. Ferruh Demirmen, formerly of the Kansas Survey and now with Royal Dutch-Shell, contributed substantially to the mathematical formulation of the project. Dr. Alfredo Prelat, formerly at Stanford University and now with the Geological Survey of Norway, gathered and interpreted large quantities of subsurface data which were essential in early stages of the project. While a visiting research scientist at Kansas, Ing. Ricardo Olea of Empresa Nacional del Petroleo (Chile) provided much of the background on regionalized variable theory, now incorporated in KOX map analysis methodology.

We are particularly indebted to Mr. James Daniels and to the Murfin Drilling Company of Wichita for generous assistance and encouragement in the application of KOX methods to the exploration for oil in northwestern Kansas. The KOX Project was directed by an advisory council which included Professor John Haun, Colorado School of Mines; Dr. Robert F. Walters, Walters Drilling Company; Dr. Paul Newendorp, John M. Campbell, Inc.; Dr. Jacques Yost and Dr. Robert Meader, Marathon Oil Company; Professor Floyd Preston, The University of Kansas; Mr. P. T. Amstutz, Petroleum Department of the Fourth National Bank of Wichita; Mr. John Stout, Petroleum Information, Inc.; Mr. Leo Broin, Mr. J. R. Pielsticker, and Mr. Robert Garner, Cities Service Oil Company; Mr. W. M. Raymond, Raymond Oil Company; and Professor Hutton Barron, The University of Kansas.

Throughout the KOX Project and the writing of this book, able secretarial and editorial assistance has been provided by Mrs. Jo Anne Kellogg, Mrs. Kathy Remark, and Mrs. Iris Troche of the Kansas Geological Survey, and by Mrs. Carol Vonder Linden of Stanford University.

Finally, the authors would like to illustrate the strength of their conviction in the utility of a probabilistic approach. All three have contributed in equal measure to the research and writing of this book. There is no senior author; the order of names on the cover was determined by the flipping of an unbiased coin!

JOHN W. HARBAUGH
JOHN H. DOVETON
JOHN C. DAVIS

Stanford, California
Lawrence, Kansas
February 1977

Acknowledgment

Acknowledgment is made for illustrations used or adapted from the following sources: American Association of Petroleum Geologists (Figures 1.4, 1.11, 1.20–1.27, 4.4, 4.5, 4.7, 5.2, 5.8–5.10, 6.1, 6.2, 6.4); *Canadian Journal of Earth Science* (Figures 1.13–1.16); Colorado School of Mines (Figures 1.10, 1.11, 1.23, 1.24, 2.7, 2.15); Harvard Business School Division of Research (Figure 8.5); *Oil and Gas Journal* (Figure 7.6); Pennsylvania State University (Figures 1.17–1.19); and Prentice-Hall, Inc. (Figure 1.12).

Contents

PROBABILITY METHODS IN OIL EXPLORATION

CHAPTER 1

The Probabilistic Nature of Oil Exploration

CHANCE AND EXPLORATION

Oil exploration is inherently probabilistic; by its very nature it includes large elements of risk and uncertainty. The circumstances that lead to the generation of oil and gas are understood only in a general sense, but certainly they must reflect the vagaries of the depositional environment. Migration of hydrocarbons into traps, and the creation of the traps themselves, are governed by processes which cannot be treated in a deterministic way except at an extremely simplistic level. The existence, or more particularly the location of traps, whether structural or stratigraphic, cannot be predicted with certainty. Even when a trap is successfully drilled, it may prove to be barren for no immediately discernible reason. Furthermore, the economic factors that ultimately affect the exploitation of resources are subject to capricious shifts which seem to defy logical prediction. Thus the entire environment of oil exploration is permeated with chance factors; so much so that the business has been referred to as "the greatest gamble on earth."

This book describes methods by which some of the aspects of chance in oil exploration can be objectively and consistently treated. Formal analysis of the pattern of chance events does not change them into certain events. It is possible, however, to establish the degree of uncertainty surrounding an event. This book is principally devoted to methods of estimating the degree of uncertainty or risk (expressed as probabilities) associated with individual oil exploration prospects.

Geologists and geophysicists have developed a variety of methods for evaluating the favorability of sedimentary basins and individual oil plays and prospects within a basin. Some of these methods are highly formalized, such as the procedures for grading prospects described by Schwade (1967), Gotautas (1963), and Benelli (1967). These provide systems for appraising the degree of favorability of a prospect in terms of a number of geological factors. They do not, however, provide objective quantitative appraisals because they involve an intuitive blending of geological qualities and quantities that have been given subjective weightings.

The industry has actively sought improved methods for rating prospects, especially methods that would minimize the element of human judgment with its attendant inconsistencies. Drilling and leasing decisions should be based on the most realistic possible appraisal of the likelihood of success, coupled with economic information and the other factors that must be considered by management. Any significant improvement in the ability to assess basins, plays, and prospects would have great benefits, and even a modest improvement in this ability should have substantial economic consequences.

Decision making under conditions of uncertainty is a special province of statistics. This mathematical field was expressly created to find ways of making decisions so that the possibility of being in error is explicitly known and at a minimum. Originally, statisticians investigated agricultural and biological concerns: combinatorial problems arising in genetics, evaluation of the influence of different treatments complicated by confusing factors, and the optimization of experimental designs. Later, statisticians developed methods of quality control: how to decide if standards were met, when to replace machine tools, how to assess the merits of alternative procedures. Most recently, statisticians have turned to significantly more difficult problems, such as attempting to characterize human behavior and predicting the economic future. The basic complexities of such undertakings are equivalent to those contained in oil exploration; the failure of statistics to provide definitive "answers" in these other areas should not discourage its application to the problems of exploration. To the contrary, it is heartening that statistical methods have shed so much light on these incredibly complex problems and that they promise to prove equally illuminating in petroleum exploration.

Statistics and probability theory have undergone enormous advances in recent decades. These advances have been strongly influenced by the advent of computers. Almost equally important, however, has been the growing realization that most problems in science, technology, and business are partly probabilistic in their nature. As a consequence, statistical methods and probability theory have been fruitfully applied to an ever-expanding variety of problems. Paradoxically, statistical tools have not received the attention they deserve in oil exploration. This is particularly surprising because oil exploration has always been regarded as an enterprise involving strong elements of chance. Furthermore, since exploration and development activities of the oil industry have been carried out over a period of decades in many parts of the world, an immense volume of data has accumulated that can be treated effectively only by statistical methods.

Uncertainty in Geologic Perception

Exploration prospects are appraised in the light of current perceptions of subsurface geology. There is an inadequate awareness of the differences between geological features as they are perceived and as they exist in reality. Understanding this difference is essential because forecasts are based on perceptions which may differ substantially from reality. Part of the difference between perceived geology and reality is due to the inadequacy of our senses, although these may be extended through devices such as logging tools which measure properties that are quite beyond our unaided comprehension. Most importantly, however, the gap between perception and reality is related to the resolution with which geologic features can be observed.

The analysis of subsurface structure, for example, is strictly limited by the available data. If the data are from wells, the number of wells and their geographic distribution are critical to our ability to interpret the structure. Of course, the actual complexity of the structure is also important. For example, if we are certain that a particular stratigraphic horizon is a perfect plane, its structure can be completely established by as few as three wells. The structure of most geologic horizons is much more complex, however, and the degree to which this complexity can be ascertained is strongly dependent on the availability of data.

Figures 1.1 to 1.3 provide an example of progressive changes in the gap between perceived and actual geology. Figure 1.1 is a structural map of an area that has been contoured using information from only 13 wells. Figures 1.2 and 1.3 show the level of detail that can be inferred from additional well information that became available at subsequent times. All three maps were made by an identical contouring procedure, so the difference between the maps is due solely to differing amounts of information.

Which of the three maps is correct? None of them, of course, because none are based on all the information that would be necessary to have a

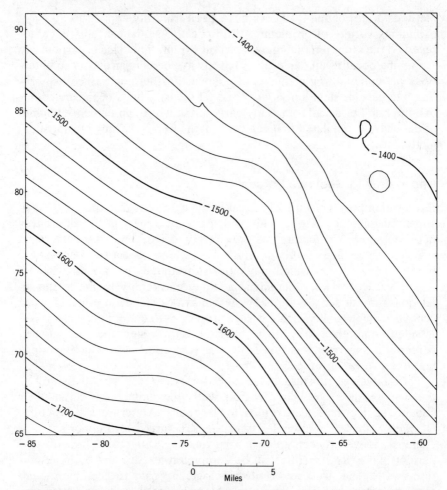

FIGURE 1.1. Structure contour map of top of Lansing-Kansas City Group (Pennsylvanian) in area in south-central Kansas. Map has been contoured by computer, using SURFACE II program developed by Sampson (1975*b*). Map is based on data from all wells that had been drilled in area before the end of 1935. Data were compiled by Prelat (1974) and contoured by Demirmen (1973*b*).

complete understanding of the structure. Complete, total perception would require a virtual infinity of wells—all measured without error. There is no doubt, however, that there is progressive improvement toward true representation of structure from the map in Figure 1.1 and that in Figure 1.3.

Figures 1.1 to 1.3 emphasize the point that our perception of geology inevitably must differ from reality in some degree. A second point is that

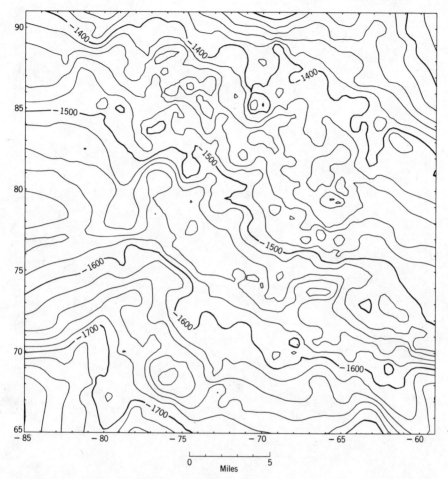

FIGURE 1.2. Structure contour map similar to Figure 1.1, except that it is based on well data available at the end of 1950.

our interpretation or perception of the geology is inherently probabilistic. For example, the elevation of any point on the contoured surface of map 1.1, except at the control points, is in some error as shown by the more recent maps. The elevations of intermediate points on the contoured surfaces shown in maps 1.2 and 1.3 are also in error, but because of the greater density of wells, the errors are generally less in map 1.2 than in 1.1, and still less in map 1.3. A succession of maps at intervening intervals would demonstrate more gradual changes with progressive reduction in errors as more well information became available.

The difference in geologic features that can be interpreted from these

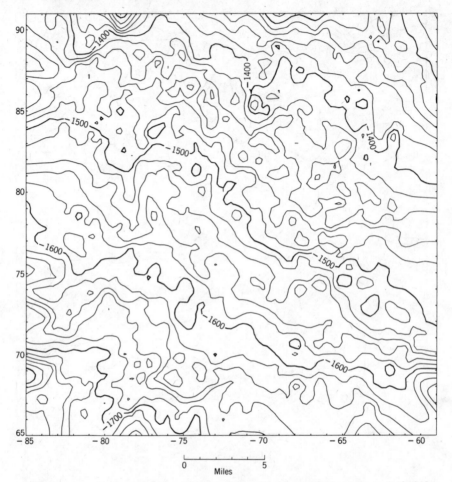

FIGURE 1.3.   Structure contour map based on well data available at the end of 1965.

maps is significant. Map 1.1, prepared from data available at that time, is based on so few wells that the structure can be objectively interpreted only as an undulating homocline which conforms to the regional dip to the southwest. By the time of map 1.2, well density has increased greatly and the perceived structure is more complex; noses, troughs, and small domes and depressions are apparent. By the time of map 1.3, the structural interpretation is still more complex.

As explorationists, our appreciation of the progressive improvement in geologic knowledge in maps 1.1 to 1.3 should be tempered by one consideration: is the increase in information relevant to the exploration objective? That is, does possession of the information in map 1.3 rather than map 1.2 improve the chances of success in wildcat drilling? Locally, it may, but a number of fields were discovered in the process of drilling the additional wells used to make map 1.3, possibly reducing the chances of additional discoveries within the area.

The sequence of maps in Figures 1.1 to 1.3 illustrates that our interpretation of structure is illusory at any stage. As more wells are drilled, knowledge of the geologic structure progressively increases up to a theoretical limit. This upper limit represents the uncertainty in interpretation of the records of individual wells. However, economic considerations invariably preclude such an advanced state of development, and our knowledge of geologic reality is uncertain primarily because of the paucity of the available data.

In addition to the uncertainty associated with interpretation of subsurface geology, we must add the uncertainty about the association between perceived geology and the occurrence of oil. To give an example, suppose that scattered wells provide enough information so that an anticline is inferred in the subsurface. The exact configuration of the anticline, however, is unclear. Many more wells would be needed to ascertain the details of its shape and size. With sufficient funds, it would be possible to drill a large number of closely spaced wells simply to gain information. Unfortunately, wells are expensive and the drilling of groups of closely spaced wells for scientific purposes alone is seldom possible. Even if this could be done some uncertainty about the exact size and shape of the structure would still remain.

Suppose, however, that we do have total knowledge of the anticline in all its geologic aspects, but no direct knowledge about the presence of oil. To what extent do changes in the uncertainty concerning the geology affect the uncertainty surrounding whether oil is present?

This difficult question appears to be raised infrequently because most petroleum geologists do not differentiate "geologic uncertainty" from "oil-occurrence uncertainty." In theory it seems that these two sources of uncertainty could be independently assessed. In practice they are difficult to separate, and their total segregation may not be possible. However, assessments of the likelihood of oil occurrence, *given* the presence of a certain geologic condition, are essential if we are to evaluate the possibility of those embarrassing drilling ventures that are "scientific successes" but "economic failures."

Elements of Probability

Probability is a numerical statement of the likelihood that an event will occur. Conventionally, probability is expressed as a percentage, or on a scale from zero (absolute impossibility) to one (absolute certainty). Probabilities can be found in various ways and may even be deduced where all circumstances and constraints are known in advance as they are in many games of chance. However, when dealing with natural events such as the discovery of oil, it is not possible to determine probabilities in this way. Rather, they must be found empirically, by observing the relative frequencies of occurrence of an event (such as the discovery of a new field) in a long series of trials (the drilling of wildcats). Eventually, the proportion of successes and failures converges to a stable estimate of the probability of success in similar experiments. The well-known "success ratio" is nothing more than a statement of the probability of success in a certain type of experiment, based on a long record of industry performance.

The idea that forecasts can be based on the record of past performance is a familiar one, because an essential element of the scientific method in geology is the comparison of areas or circumstances with similar areas or circumstances that have already been studied. Probability is therefore nothing more than quantified experience, and the use of probability statements in the evaluation of prospects is simply an extension of the process used in the industry in a less formal and more subjective way.

However, there are several significant advantages in recasting geological "experience" into a probability framework. The first of these is that it requires a rigorous definition of the conditions of a "trial," to ensure that the relative frequencies really reflect the proportions of *similar* events. This forces a close examination of the record of prior occurrences, which can be an instructive experience in itself. Second, it places all prior "experiences" on a common scale so they can easily be compared and ranked. Similarly, the common scale for probability of success allows comparison between competitive alternatives, even though the circumstances surrounding these alternatives are markedly dissimilar. Most importantly, use of an explicitly quantitative expression of the probability of success in an exploration venture allows us to use the powerful tools of econometric analysis. In turn, predictions may be made of the financial benefits (or penalties) that are the most likely result from a certain course of action. Econometric forecasts are used to guide almost all major corporate activities, including investment policy, capital expansion, and the like. Although these methods are invaluable in production, refining, and marketing segments of the petroleum industry, they have not been as effectively used in exploration because of the lack of information about the probability of success attached to individual drilling decisions.

If two events are possible outcomes of a trial and they cannot occur simultaneously, they are said to be *mutually exclusive*. The drilling of a wildcat well is a trial, and a dry hole or the discovery of oil are possible outcomes that are mutually exclusive. The probability that one event *or* the other happens is the sum of their separate probabilities: P(oil or dry) $= P$(oil) $+ P$(dry). This is the *additive rule of probability*.

If two events are not mutually exclusive but are independent of one another, then the joint probability that they will both occur simultaneously is the product of their separate probabilities of occurrence. For example, suppose P(anticline) is the probability that a trial well penetrates an anticlinal structure, and P(sandstone) is the probability that a particular formation encountered is a sandstone. The joint probability that both an anticline and a sandstone are encountered by the drill is P(anticline and sandstone) $= P$(anticline) $\times P$(sandstone). This is the *multiplicative rule of probability*.

Now, if the occurrence of two events is dependent to some degree, their joint probability of occurrence is *conditional*. This is most important, because we are primarily interested in those probabilities of oil occurrence that are conditional on some geologic circumstance. If a conditional relationship exists, then the geology provides information about the likelihood of oil. If P(oil) is the probability of a successful well and P(anticline) is the probability of an anticlinal feature, then P(oil and anticline) $\neq P$(oil) $\times P$(anticline). The conditional probability that oil occurs, *given* that an anticline has been found, is designated P(oil| anticline).

In the remainder of this book, we are concerned with many aspects of probability as they pertain to the likelihood of the discovery of oil. For example, we are concerned about the conditional relationships between oil occurrence and the presence of certain types of structures, stratigraphic units, or other geologic properties. We also are interested in the probability that the geologic features which we perceive from our limited data correspond to geologic reality. These probabilities are combined in various ways to estimate a final probability of success that can be used to evaluate a specific prospect. Throughout these pages we use words such as likelihood, error, and uncertainty. These all relate to probability; either the probability of something (usually oil) being present, or the probability that what we perceive is a valid expression of reality.

Alternatives in Probabilistic Exploration

Although probability theory has not been widely applied in petroleum exploration (or indeed in any area of geology), research to date has followed several contrasting approaches. These fall roughly into three broad categories: basin evaluation, search strategy, and conditional analysis. Of these, basin evaluation is the most developed, in part because it is the easiest and because its objectives are more limited than those of the other two approaches.

Basin analysis procedures attempt to predict the amount of recoverable oil in a sedimentary basin or major petroleum province. These possible resources are estimated from studies of more mature basins believed to be geologically similar, and by evaluating the statistics of oil fields discovered in the basin. Important clues to the remaining potential of a basin are gained from the sizes of fields discovered, the drilling success ratio, and similar measures that can be extrapolated into the near future. Although basin analysis provides valuable comparative estimates useful for long-range planning, both by industry and government, it does not address the far more difficult problem of *locating* the undiscovered resources within a basin.

Search strategies include various schemes for systematically sampling a prospective area. They include grid-drilling programs and more prosaic methods such as drilling as close as possible to proven production. An entire body of statistical theory is devoted to the problem of hitting a hidden target of specified size and shape in the minimum number of attempts. These methods usually require assumptions about the physical dimensions and number of potential targets and may require assumptions about the statistical nature of their spatial distribution as well. The necessary information is gained from an examination of more mature areas where these parameters can be determined. Although much valuable information has been gained from such studies about the spatial distribution of mineral deposits, including oil fields, they have been employed only to a very limited extent in the oil industry. John C. Griffiths of Penn State University and his associates have been leading proponents of grid drilling over large areas. In the highly competitive environment of free enterprise in the United States, grid-drilling strategies would of necessity be confined to the exploration of large tracts such as those offshore. Concessions extending over vast areas in sparsely explored countries, however, might be particularly well suited for grid drilling.

Conditional analysis attempts to assess the probabilistic relationship between geologic variables and the occurrence of petroleum. Because the state of these geologic variables is usually easier to observe than oil itself, they provide indicators to the presence of the exploration target. This is exactly the approach used by traditional explorationists, except that the relationship between the guiding variables and oil is in the form of explicitly expressed probabilities. These probabilities are further modified by the uncertainty in our perception of the geologic variables themselves. This uncertainty is related to the density of available control, which may be in the form of wells, seismic lines, or other observations.

Again, the conditional probabilities may be derived by examining the historical development of a mature area which is considered to be geologically similar to the exploration area. However, a large number of geologic variables may be assessed as potential guides for exploration. The relative efficacy of these different geologic properties can be evaluated, and the most efficient ones combined into a final search tool. The evaluation process does not stop, however, as the conditional probabilities constantly change as experience is gained during the exploration program. Conditional analysis is a systematic attempt to mimic the learning process by which an explorationist progressively gains an understanding of an oil province. Of course, there *is* a possibility that no conditional relationship can be established between geologic variables and oil occurrence in a prospective area. This would not only thwart a probabilistic exploration strategy based on geologic relationships, but would also deny geologic analysis of any sort. As geologists we find it hard to accept the existence of such circumstances.

Unfortunately, an understanding of the geological controls of many oil accumulations has emerged only as rationalizations after the initial discoveries were made. There are only hazy recollections of the roles, if any, that the geological perceptions had in leading to the discoveries. The vast majority of oil fields described in the geological literature provides little assistance on this issue, because almost invariably these descriptions are based on interpretations from wells drilled after discovery of the fields.

PROBABILITY AND BAYESIAN ANALYSIS

In the previous pages, we have dwelled on some of the means for obtaining estimates of probabilities for exploration. Bayes' theorem provides another way in which frequency data can be used to calculate probabilities that depend on knowledge of previous events. Suppose there exists a set of mutually exclusive and exhaustive events that are considered possible. It is known in advance that only one of these events will actually occur, but it is uncertain as to which event this will be. Initially, a probability can be assigned to each of these events on the basis of available evidence or informed judgment. If additional evidence later becomes available, the initial probabilities can be revised by means of Bayes' theorem which describes the relationship between *joint, marginal,* and *conditional probabilities*. There is a large literature on "Bayesian" methods in statistical decision processes, such as the introductory monograph by Morgan (1968) and the book by Martin (1967).

Dowds has been one of the principal proponents of Bayesian analysis in petroleum exploration. In a series of papers (Dowds, 1961, 1964a, 1964b, 1965, 1966, 1968, 1969a, 1969b, and 1972), he has shown that conditional probabilities based on earlier available information may be progressively readjusted as additional drilling information becomes available. The probability estimates can be presented as a series of contour maps that forecast the success ratios of exploratory holes to be drilled.

Several papers have also appeared that deal with the use of Bayesian probabilities in a mineral exploration context. Brock (1974) elicited responses from various geologists concerning their subjective appraisal of geologic factors influencing the occurrence of porphyry copper deposits in Arizona. The respondents assigned probabilities to sets of geological variables, and in turn these were combined as a series of branching probability "trees." An earlier study by Harris, Freyman, and Barry (1970) employed subjective estimates of probabilities in an appraisal of the mineral potential in northern British Columbia and in the Yukon Territory.

Rather than provide formulas for Bayes' theorem in its various guises, we illustrate Bayesian concepts here with a simple hypothetical example. Table 1.1 contains hypothetical frequencies listed by three kinds of geologic structure: anticlines, homoclines, and synclines, which are mutually exclusive. The structures either have faults or lack faults, categories which are also mutually exclusive. Finally, the structures are either productive or dry. Thus, there are three sets of qualities, and within each set the particular categories are mutually exclusive. These frequencies can be transformed into probability estimates by dividing by appropriate totals. Table 1.2 gives probabilities that are conditional on the specified geologic structures.

Either table can be used to estimate the conditional probabilities in a

TABLE 1.1. Frequency data pertaining to hypothetical oil-field structures. Wells are subdivided as producing or dry (italics).

	Without Faults (W)			With Faults (F)			
	Pro-ducing (G)	Dry (D)	Sub-total	Pro-ducing (G)	Dry (D)	Sub-total	Row Totals
Anticline (A)	200	300	500	60	40	100	600
Homocline (H)	10	40	50	75	175	250	300
Syncline (S)	5	20	25	25	50	75	100
Column totals	215	360	575	160	265	425	1000

TABLE 1.2. Probabilities associated with hypothetical oil-field structures.

	Without Fault (W)			With Fault (F)			Marginal Probabilities According to Fold Type
	Producing (G)	Dry (D)	Probability	Producing (G)	Dry (D)	Probability	
Anticlines (A)	0.200	0.300	P(A,W) 0.500	0.060	0.040	P(A,F) 0.100	P(A) 0.600
Homoclines (H)	0.010	0.040	P(H,W) 0.050	0.075	0.175	P(H,F) 0.250	P(H) 0.300
Synclines (S)	0.005	0.020	P(S,W) 0.025	0.025	0.050	P(S,F) 0.075	P(S) 0.100
Marginal probabilities according to presence of faults	P(G,W) 0.215	P(D,W) 0.360	P(W) 0.575	P(G,F) 0.160	P(D,F) 0.265	P(F) 0.425	1.000

variety of situations. Consider the following hypothetical questions and responses:

1. A fault is detectable at the surface, but there is no other knowledge about the subsurface structure. What is the probability of a dry hole?

$$P(D|F) = \frac{P(D, F)}{P(F)} = \frac{0.265}{0.425} = 0.624$$

2. No geologic information is available. What is the probability of a dry hole?

$$P(D) = \frac{P(D, W) + P(D, F)}{1.000} = \frac{0.360 + 0.265}{1.000} = 0.625$$

3. (a) A well is begun on a structure known to be an anticline. What is the probability of a producer?

$$P(G|A) = \frac{0.200 + 0.060}{0.600} = 0.433$$

(b) As the same well is being drilled, fault gouge is encountered. What is the probability of a producer conditional on the fact that a fault is present and the fact that the structure is an anticline?

$$P(G|A, F) = \frac{0.060}{0.100} = 0.600$$

4. A well is being drilled with no initial knowledge of the structure. A fault is later encountered while drilling. Given knowledge that a fault is present, compute the probabilities that the structure is (a) an anticline, (b) a homocline, and (c) a syncline:

$$P(A|F) = \frac{P(A, F)}{P(A, F) + P(H, F) + P(S, F)} = \frac{0.100}{0.425} = 0.235$$

$$P(H|F) = \frac{0.250}{0.425} = 0.588$$

$$P(S|F) = \frac{0.075}{0.425} = 0.176$$

5. Given the knowledge that a fault is present, but no other knowledge about the structure, what is the probability of a producer?

$$P(G|F) = \frac{P(G, F)}{P(G, F) + P(D, F)} = \frac{0.160}{0.160 + 0.265} = 0.376$$

These examples illustrate that Bayesian methods are simple and yet powerful. It is surprising that Bayesian methods have been used so seldom in oil exploration; possibly this is because little effort has been devoted to the systematic tabulation of frequencies required for their application.

Combining Empirical Probabilities Through Monte Carlo Summation

Forecasts of oil and gas that remain to be discovered tend to involve a composite mixture of both "known reserves" and "unknown resources." In fact, a spectrum of degree of uncertainty can be established, extending from "measured" through "indicated," "inferred," "hypothetical," and finally to "speculative" resources. These categories are part of a continuum that grades from certainty at one end to extreme uncertainty at the other end. A problem arises when universal estimates—which are necessarily probabilistic—are to be combined to yield a gross estimate. Suppose, for example, that we have three categories, which in order of decreasing certainty are termed "probable," "possible," and "speculative." In each category, there is a range from a minimum through the median to the maximum likely value. Since the three categories are mutually exclusive, it is apparent that they should be summed together somehow to yield overall estimates. But simple addition of the minimums and maximums is not appropriate because the range within each category is also a continuum. The solution is to use Monte Carlo methods for their summation, which will yield a continuum that can be represented by a curve.

The following example is adapted from the work of White, Garrett, Marsh, Bates, and Gehman (1975). Figure 1.4 contains three probability

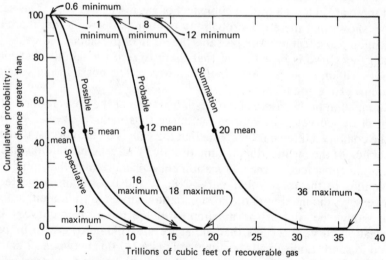

FIGURE 1.4. Producible gas in south Louisiana estimated in late 1974. Areas to left of *speculative, possible*, and *probable* curves have been summed to yield curve labeled *summation*. Adapted from White and others (1975).

curves which represent the probable, possible, and speculative supply of gas in south Louisiana as estimated by a research group at Exxon in 1974. The fourth curve represents their sum.

Probable supply is defined as an extension of data from existing fields in south Louisiana. The probable supply includes likely future extensions of existing fields, revisions of estimates, discovery of new pools within existing fields, and expected secondary recovery developments that are based on past experience in other secondary recovery projects. The *possible* supply may result from relatively new fields in volumes of sedimentary rocks that have only been partially explored by drilling. The possible supply has been estimated by extrapolation of historical rates of discovery in south Louisiana and in other regions. The *speculative* supply consists of gas that may come from volumes of sedimentary rocks that are inferred to be present but have not yet been drilled with sufficient density to locate major gas fields. The speculative supply has been estimated by comparison of the inferred character and volume of the sedimentary rocks in question with other sedimentary rocks that are of similar character, but which have been extensively explored.

Our quest here is not with the details of the assessment by White and others, but rather the methodology by which the estimates are combined. Each curve of Figure 1.4 represents the estimates as "greater than" probabilities at each point along the curve. For example, the probable curve shows that there is 100 percent probability (i.e., absolute certainty) that there are at least 8 trillion cubic feet of gas in this category. This same curve shows that there is a probability of about 45 percent that there are 12 trillion cubic feet or more present (the mean of the calculated values on which this curve is based), and that there is zero probability that more than 18 trillion cubic feet exist in this category. The possible and speculative curves are read similarly.

Summation of the curves involves integration of the areas to the left of each of the curves. The summation curve thus includes the same areas as the combined three curves. Note that the values on the curves, with the exception of the means, do not sum directly. Thus, the minimum value of 12 trillion cubic feet on the summation curve is more than the sum of the minimums of the three other curves (8, 1, and 0.6 trillion cubic feet). Conversely, the maximum of the summation curve (36 trillion cubic feet) is less than the sum of the maximums of the three other curves. The summation curve forecasts that the aggregate volume of gas to be produced in south Louisiana (as estimated in late 1974) is between 12 and 36 trillion cubic feet, with a continuum of probabilities in this range.

The summation process can be effectively carried out by a Monte Carlo procedure. Since the curves themselves are empirical, they have no analytical function that can be integrated and evaluated between limits to yield the area to the left of the curve. Instead, a large number of points (perhaps 25 to 50 thousand) can be chosen at random along the three curves and subsequently combined by addition. Such an operation would be prohibitive if attempted manually, but can be performed in a few seconds by a large computer.

The United States Geological Survey (USGS) has employed Monte Carlo methods for petroleum resource appraisal for the entire United States, on a province-by-province basis (Miller and others, 1975). Figure 1.5 illustrates the general classification of petroleum resources according to the degree to which they are demonstrated to exist based on geological information (the degree of "geologic assurance"), and the degree of economic feasibility of their exploitation. Petroleum resources in the United States form a continuum with respect to degree of assurance and degree of feasibility, but for estimation purposes it is essential to split them into discrete categories as in Figure 1.5.

Geologists with the USGS subjectively estimated the volume in barrels of oil or trillions of cubic feet of gas in the various categories. Estimates were made of in-place oil and gas, total recoverable resources, and remaining undiscovered recoverable resources, as follows:

1. A low resource estimate corresponding to a 95 percent probability that there is *at least* that amount.

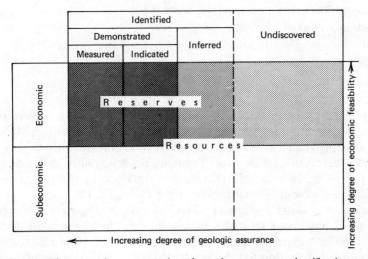

FIGURE 1.5. Diagrammatic representation of petroleum resource classification employed by the U. S. Geological Survey and the U. S. Bureau of Mines (Miller and others, 1975).

2. A high resource estimate with a 5 percent probability that there is *at least* that amount.

3. A modal estimate of the resource which the estimator associates with the highest probability of occurrence that there will be that amount.

4. A statistical mean that is calculated by adding the low value, the high value, and the modal value and dividing the sum by 3.

These estimates are translated into United States resource assessments by aggregating all provinces. This involves Monte Carlo summation of the individual distributions of the provinces which are assumed to be statistically independent. Lognormal reserve distributions were assumed for most areas. There are regions in the United States where no petroleum has been discovered. Consequently, there is uncertainty as to whether commercial quantities of oil and gas exist at all in these areas. Therefore, a marginal probability was assigned to the events "commercial oil found" and "commercial gas found." These marginal probabilities were subjectively set by consensus of a committee on a province-by-province basis. In turn, the marginal probabilities have been combined to yield an estimate for the whole of a region embracing several provinces. The manner of combining the estimates is illustrated below. The central onshore Alaska region consists of four provinces, for which the following estimates of the probability of undiscovered, commercial oil *not* being present are:

1. Yukon-Porcupine province: 70 percent.

2. Yukon-Koyukuk province: 75 percent.

3. Interior Lowlands province: (negligible amounts; probability not estimated).

4. Bristol Bay Tertiary province: 60 percent.

The probability of undiscovered oil not being present in the region is obtained by multiplying the estimates for the three provinces (the Interior Lowlands province being omitted), which yields the quantity $0.70 \times 0.75 \times 0.60 = 0.315$, or 31.5 percent. The marginal probability of finding oil is thus $1.00 - 0.315 = 0.685$. This figure forms one of the points on the cumulative probability distribution curve of Figure 1.6.

The other onshore regions of Alaska (northern, Figure 1.7; southern, Figure 1.8) have been treated similarly. In turn they have been summed by Monte Carlo methods to yield an estimate which is also in the form of a cumulative distribution for the whole of onshore Alaska (Figure 1.9).

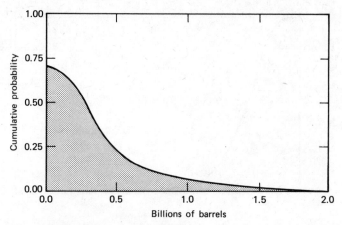

FIGURE 1.6. Cumulative probability distribution of estimated undiscovered recoverable oil resources in onshore central Alaska. Adapted from Miller and others (1975).

CHARACTERISTICS OF OIL-FIELD POPULATIONS IN PETROLEUM BASINS

Design of an efficient exploration strategy requires a detailed understanding of the relative sizes, shapes, frequencies of occurrence, and spatial arrangements of the oil fields which are the targets of wildcat drilling. In the initial exploration of a virgin area, these statistics are unknown but

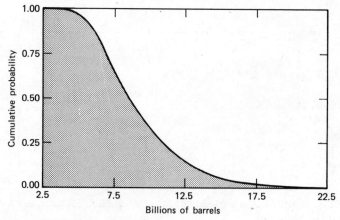

FIGURE 1.7. Cumulative probability curve of estimated, undiscovered recoverable oil resources in onshore northern Alaska. Adapted from Miller and others (1975).

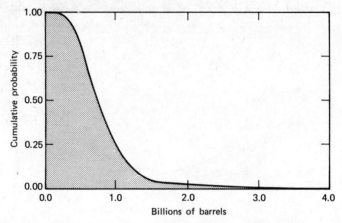

FIGURE 1.8. Cumulative probability curve for onshore southern Alaska. Adapted from Miller and others (1975).

their general characteristics are probably similar to those in other petroleum basins.

Hubbert (1967), in a classic paper dealing with the degree to which the United States has been explored for petroleum, used graphs of the annual volumes of production for the country as a whole. These graphs tend to assume the form of a normal or Gaussian curve, and in certain aspects they can be analyzed from a probabilistic viewpoint. Frequencies of the volumes of oil fields within any individual basin generally follow a log-

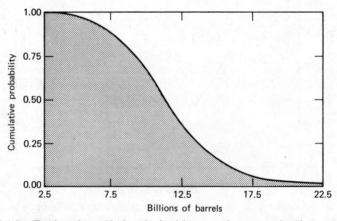

FIGURE 1.9. Total onshore Alaska obtained by summing curves in Figures 1.6 to 1.8. Adapted from Miller and others (1975).

normal distribution. This is true, for example, in the Denver basin (Arps and Roberts, 1958; Haun, 1971), in north Louisiana (Kaufman, 1963), and in Kansas (Griffiths, 1966). A variable is lognormally distributed when the logarithms of the observed values form a normal distribution, which plots as the familiar bell-shaped curve. When graphed as cumulative percentage on log probability paper, a distribution is lognormal if the observations fall on a straight line. These graphical properties are illustrated by oil-field volumes from the Denver basin (Haun, 1971) shown in Figures 1.10 and 1.11.

The lognormal distribution is widely used in geology to characterize sediment size variations (Krumbein and Pettijohn, 1938) and geochemical element concentrations (Ahrens, 1954). The normal distribution is applicable in situations where observations are symmetrically arranged about a central value which coincides with the mean, and both small and large values are relatively rare. In contrast, the lognormal distribution shows a strong positive skew with a "tail" of small counts of large values. At the other extreme, there is a relatively high frequency of low values. This model matches the explorationist's experience in most basins, where the comparatively rare giant fields are greatly outnumbered by a host of small fields. As a theoretical model, the lognormal distribution is generally the result of a multiplicative process, generated by the random splitting of larger entities into more numerous smaller entities (Aitchison and Brown, 1969). This contrasts with the normal distribution model where the spread in variation about a central "expected" value is the cumulative result of

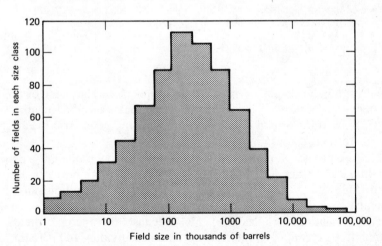

FIGURE 1.10. Histogram of oil-field sizes in Denver basin at end of 1969. From Haun (1971).

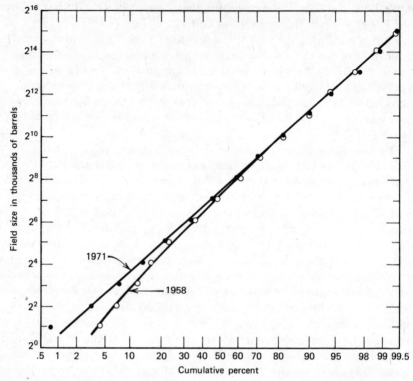

FIGURE 1.11. Frequency distributions of oil-field sizes in Denver basin, based on fields discovered through 1958 (open circles) and through 1971 (solid circles). Distributions have been graphed as cumulative percentages on log probability paper. Distribution of fields based on 1971 data more closely approximates an ideal lognormal distribution than distribution based on 1958 data. Adapted from Haun (1971) and Arps and Roberts (1958).

small arithmetic displacements. The multiplicative character of the lognormal distribution is illustrated by the fact that the mean of the logarithmic transformed data coincides with the geometric mean of the untransformed values.

The technique of plotting a distribution on log probability paper is illustrated in Figure 1.12 and Table 1.3. The data pertain to total ultimate producible oil by field in millions of barrels in north Louisiana as of January 1946, based on fields containing one million barrels or more.

The tabulation includes 25 fields. Because there are 25 fields, the cumulative percentage of fields is obtained by dividing 100 percent by 25 + 1 to obtain the fractional percentage to be assigned to each field. This is essential in using logarithmic probability paper, which contains neither 0

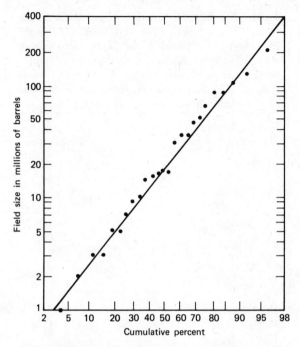

FIGURE 1.12. Cumulative frequency distribution of ultimately producible oil volumes of north Louisiana fields containing one million barrels or more as of January 1946. Points represent volume per individual field (vertical) plotted against cumulative percentage (horizontal). Data are expressed numerically in Table 1.3. From Kaufman (1963).

nor 100 percent on the cumulative probability scale. The "fractile" percentage devoted to each oil field in this example is 100/26 = 3.846. Also, the oil fields must be sorted into a sequence by size. As the graph (Figure 1.12) reveals, even at this early or intermediate stage in the history of north Louisiana's oil-field development a lognormal distribution of field sizes is apparent. Subsequent tabulations of oil field sizes in north Louisiana adhere to the lognormal distribution.

Probability estimates can be readily based on a lognormal distribution. If such a distribution is projected as a forecast, the probabilities for discoveries of various magnitudes can be read directly from the graph. For example, Figure 1.12 shows that there is a 65 percent probability that any randomly selected large field in north Louisiana will be greater than 10 million barrels and a 35 percent probability that it will be less than 10 million barrels. The correspondence between several field sizes and probability estimates based on the 1946 north Louisiana data is presented in

TABLE 1.3. Total ultimate producible oil in fields containing one million or more barrels as of January 1946 in north Louisiana. Left column contains magnitude of individual fields rounded to nearest million barrels, arranged in increasing size. Right column contains cumulative percentage fraction assigned to each field. Modified from Kaufman (1963).

Oil-field Size in Millions of Barrels	Cumulative Percentage
1	3.85
2	7.69
3	11.54
3	15.38
5	19.23
5	23.08
7	26.92
9	30.77
10	34.62
14	38.46
15	42.31
16	46.15
17	50.00
17	53.85
30	57.69
35	61.54
35	65.38
45	69.23
50	73.08
65	76.92
85	80.77
85	84.62
105	88.46
124	92.31
200	96.15

Table 1.4. This table does not pertain to wildcat well probabilities, but only to discoveries of fields containing one million barrels or more.

Perturbations of the simple lognormal distribution model may arise in provinces where there are several distinct "families" of oil or gas fields. An example is provided by gas field volumes in western Canada, whose bimodal frequency distribution deviates markedly from a lognormal shape. The form of this distribution has been interpreted by McCrossan (1969) as the product of two genetically distinct populations of gas fields

TABLE 1.4.   Probabilities estimated from lognormal frequency distribution of oil fields greater than one million barrels in north Louisiana on January 1, 1946.

Field Sizes in Millions of Barrels	Probability That Field Will Be Greater Than This Size (%)	Probability That Field Will Be Less Than This Size (%)
5	80	20
10	65	35
20	48	52
50	27	73
100	14	86
200	6	94

which differ in statistical properties and occur under different geological conditions. When the two populations are segregated, their separate frequency distributions approximately conform to the lognormal model (Figures 1.13 to 1.15). It seems likely that other basins containing several distinct populations of hydrocarbon reservoirs are characterized by a compound distribution of field sizes of which each population is essentially lognormal.

The importance of segregating distinct populations is graphically em-

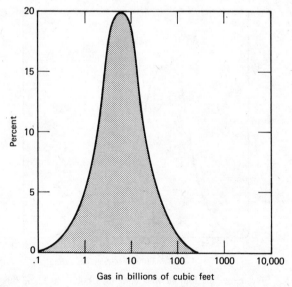

FIGURE 1.13.   Frequency distribution of small gas fields in western Canada, based on 1965 data. From McCrossan (1969).

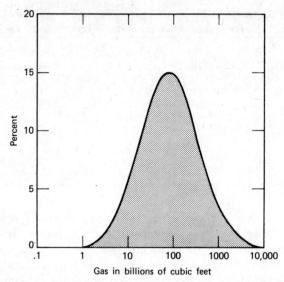

FIGURE 1.14. Frequency distribution of large gas fields in western Canada, based on 1965 data. From McCrossan (1969).

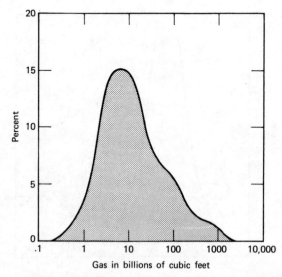

FIGURE 1.15. Frequency distribution that combines both small and large gas fields in western Canada. Curve is composite of curves in Figures 1.13 and 1.14. From McCrossan (1969).

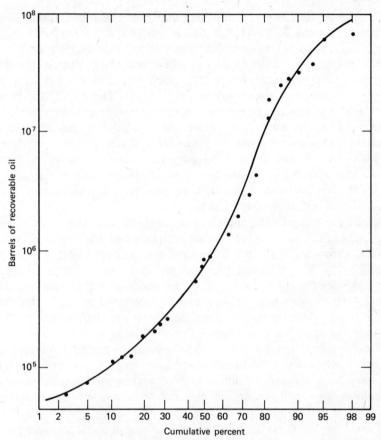

FIGURE 1.16.   Cumulative distribution of volumes of recoverable oil in Viking Sand fields in Alberta, based on 1969 information. From McCrossan (1969).

phasized in Figure 1.16, which represents the cumulative distribution of ultimately recoverable oil in fields in the Cretaceous Viking Sand of Alberta. The fitted curved line is an example of a strongly bimodal distribution. McCrossan interprets the distribution as a composite of two overlapping lognormal populations. One population consists of smaller fields and the other of the larger fields. The population of smaller fields has a geometric mean of about 0.3 million barrels and forms about 60 percent of the total number of Viking oil fields. The population of large fields has a geometric mean of about 12.6 million barrels and includes about 40 percent of the total oil fields in the Viking Sand. An alternative interpretation is that large Viking fields should not be segregated from the small fields as they both occur under similar geologic conditions. Under this interpretation, however, the total population of Viking fields forms a distribution that is more complex than a lognormal distribution.

The statistics on oil-field volumes allow us to make statements about the probability of finding fields of various sizes, by computing simple frequency ratios from the observed distributions. The "real" distribution of oil-field volumes can only be known completely when every field in a basin has been discovered, by which time probability statements are of purely academic interest. However, inferential statistics allow us to esti-mate the distribution of the total population from the relatively small sample provided by fields discovered in an early phase of exploration. Naturally, the confidence that can be attached to such estimates is a function of the size of the initial samples.

Forecasts of the total population of oil fields in a basin are of great practical and commercial importance, as shown by the strong interest in regional estimates of undiscovered oil and gas resources (Miller and others, 1975). It is reasonable to assume that field volumes in regions where exploration is beginning will be approximately lognormally distrib-uted. If a maximum field volume can be estimated with any degree of reliability, the spectrum of smaller field volumes can be predicted by graphing the cumulative frequency distribution as a straight line on log probability paper. The lower limit to field volumes can be assumed as the size below which it is uneconomic to produce. Even though estimating the maximum field size may be difficult, the tendency for the sizes of oil and gas fields to approach a lognormal distribution on a regional basis can be a useful guide in exploration planning.

In graphing the hypothetical lognormal distribution, not only must the size of the largest field and the smallest field be assumed, but the approx-imate number of fields must also be assumed. For example, if about 100 fields are assumed to be present, it would be appropriate to plot the maximum field size at the 99 percentile mark, and the smallest field size at the one percentile mark. If a total of about 500 fields is assumed, the maximum and minimum field sizes should be plotted at the 99.8 and 0.2 percentile marks, respectively. Connecting the two points thus plotted with a straight line yields a graph of an ideal lognormal distribution, from which probability estimates concerning different field sizes can be made.

Studies by Drew and Griffiths (1965) in West Texas, the Denver basin, and Indiana show that other measures of the size of oil and gas fields tend to be lognormally distributed, including their lengths and areas. De-viations from lognormality probably result from heterogeneity in the populations of fields and statistical fluctuations reflecting the relatively small size of the samples. The shapes of fields, as expressed by the ratio of length divided by width, were found to have an approximately normal distribution. Graphs of field sizes and shapes versus the dollar value of oil and gas produced (permitting oil fields and gas fields to be combined) were made for the three regions and are shown in Figures 1.17 to 1.19. Dollar equivalents of $2.90 per barrel of oil and 15¢ per thousand cubic feet of gas were assumed, as these prices were representative at the time the study was made. Reserves remaining to be produced were not included in the study.

Drew (1972) has analyzed the frequency distributions of oil-field sizes in Kansas on an areal basis. His work was motivated by the realization that explorationists generally search for productive acreage and not for individual oil fields. Typically, exploratory efforts are confined to particular lease blocks or to those areas potentially leasable.

Drew divided Kansas into a series of square cells each two miles on a side, yielding 17,280 cells for the entire state. He then tabulated the frequencies of occurrence of the various sizes of oil fields in these cells. The resulting frequency distributions do not correspond to a lognormal distribution but change as a function of cell size. Although a lognormal

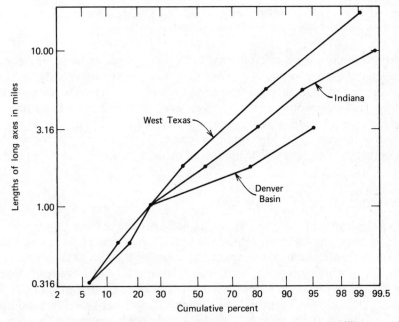

FIGURE 1.17. Cumulative frequency distributions plotted on log probability paper for lengths of long axes of oil fields in three regions. Populations involve 104 fields in West Texas, 298 fields in Denver basin, and 180 fields in Indiana. From Drew and Griffiths (1965).

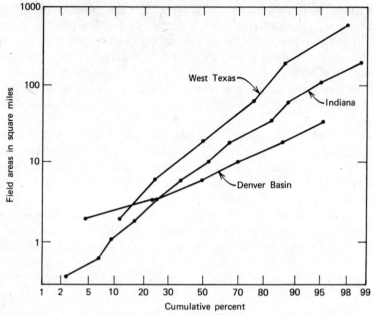

FIGURE 1.18.   Cumulative frequency distributions of oil and gas field areas in West Texas, Denver basin, and Indiana. From Drew and Griffiths (1965).

distribution is often a suitable model of oil-field sizes within a province, it may not be appropriate for areas of a few square miles. The serious disparity between an ideal lognormal distribution and the distribution found by Drew would be emphasized even more for most lease blocks, since these commonly are less than a square mile in area in Kansas and in much of the rest of the Midcontinent region. Additional research is needed to obtain the frequency distributions of oil occurrence that pertain to such small areas.

Sampling Oil-Field Populations

If a sample of oil-field sizes is to be a good estimate of the total population it must be a "random sample" in which every individual field has an equal chance of being included. Exploration may be thought of as a selection procedure which gradually accumulates a sample of discoveries from an initially unknown population of oil fields. If wells were drilled at randomly selected locations (or, indeed, according to any other search procedure), larger fields would tend to be found early simply because they are bigger targets. This pattern of discovery is borne out by the history of many petroleum provinces. Large fields are discovered early during an inten-

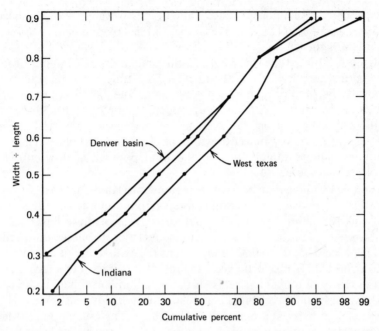

FIGURE 1.19. Cumulative frequency distributions of shapes of oil and gas fields in West Texas, Denver basin, and Indiana. From Drew and Griffiths (1965).

sive exploration phase; late discoveries tend to be small in size. The result of this bias in the selection procedure is a progressive change in the frequency distribution of field sizes through time. If the population distribution is lognormal, samples at different times also will tend to be lognormal. However, these distributions will migrate downward from an initial distribution that has an unrealistically high mean caused by inclusion of a disproportionate number of large fields. When ranked in order of discovery these sample distributions should gradually converge on the true population mean.

These general points are illustrated by Kaufman and others (1975) who simulated the discovery of fields in a hypothetical basin by a series of Monte Carlo experiments. Starting with a population of undiscovered fields whose volumes were lognormally distributed, Kaufman ran a series of trials in which fields were successively "discovered" by random selection. As each field was found it was withdrawn from the population so that there was progressive reduction in the remaining set of undiscovered fields. The probability that any particular field would be found was a function of its size relative to the sum of the sizes of fields that remained to be discovered. As a consequence, larger fields were more easily discovered than small fields and the average size of discovered fields was biased towards a high value in small samples.

Figure 1.20 illustrates a Monte Carlo simulation of sampling from a hypothetical population of 600 oil fields. The straight line represents the lognormal distribution of the total population. The curves are cumulative frequency distributions of randomly selected samples of 10, 30, 50, and 150 fields, computed as the averages of many sampling trials. Small samples are markedly biased towards larger fields, but the sample distributions migrate rapidly toward the total population distribution as sample size is increased.

A simple example may be used to demonstrate Kaufman's Monte Carlo procedure. Suppose a population consists of only four fields with sizes 1, 3, 6, and 10 million barrels. On each draw, one of these fields will be "discovered." On the first draw the discovery probabilities attached to each field are 0.05, 0.15, 0.30, and 0.50, as the probability of discovery of a particular field is proportional to the ratio of its size to the aggregate of sizes of the total remaining fields. If the 6 million barrel field happens to be drawn first, the probabilities attached to the fields that remain (1, 3, and 10 million barrels) are 0.07, 0.22, and 0.71. The assumption made in Kaufman's simulation process favors the early discovery of large fields.

Kaufman's experiments also consider the effects of different total population sizes. For example, if the total population of fields is only 100, the bias toward the discovery of large fields early in the exploration history is less (Figure 1.21) than if the discoveries are made from a total population of 600 fields. The frequency distributions of fields that remain to be discovered are also markedly different, as comparison of Figure 1.21 reveals. Obviously, it would be useful in exploration planning to be able to estimate the frequency distribution of fields remaining to be discovered. An immediate question relates to whether the frequency distribution of the total population of oil and gas fields in a basin can be forecast from the fields discovered at an early or intermediate stage of exploration. While such predictions are possible in theory, the practical difficulties are great. Because the frequency distribution of early discoveries is displaced more toward large field sizes if these early discoveries are drawn from a large total population, the rate at which the frequency distribution is progressively displaced with continued exploration could be used as a basic key to total population size. A difficulty is that the historical development of a given oil-producing region is but a single "trial" in a Monte Carlo sense.

An alternative is to treat the frequency distribution of sizes of a field discovered at some particular time in the sequence of discovery, as for

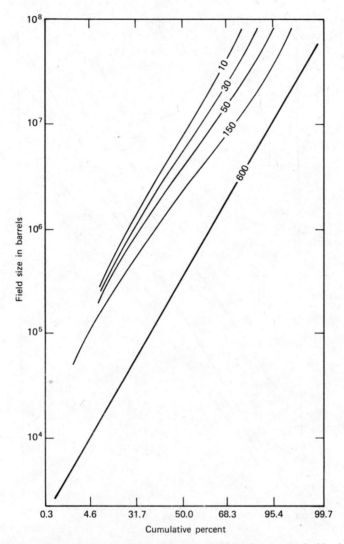

FIGURE 1.20. Monte Carlo simulation of discovery of fields in hypothetical basin containing 600 fields. Heavy straight line represents cumulative distribution of total population. Curves represent cumulative distributions of populations for different numbers of fields "discovered." Adapted from Kaufman and others (1975).

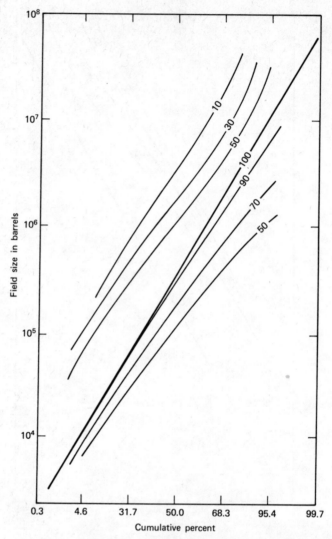

FIGURE 1.21. Hypothetical cumulative frequency distributions of sizes of oil fields discovered and remaining to be discovered, drawn from population of 100 oil fields that have a lognormal size distribution. Graphs are averages of 1000 Monte Carlo sampling experiments. Total population has a median field size of 403,400 barrels, and a mean field size of 1,808,000 barrels. Heavy straight line represents distribution of total population. Curves above straight line represent distributions of various numbers of fields discovered. Curves below straight line represent corresponding numbers of fields that remain undiscovered. Adapted from Kaufman and others (1975).

example, the 40th field. If fields are drawn from a large total population, the distribution of the sizes of the 40th discovery will be displaced more toward the large field sizes than if the fields are from a small total population. Figure 1.22 shows cumulative frequency distributions derived by Monte Carlo methods for the 40th field discovered in total populations of 100, 150, 300, 600, and 1200 fields, respectively.

Monte Carlo experiments can incorporate a variety of different assumptions. While these experiments have provided valuable insights, they are

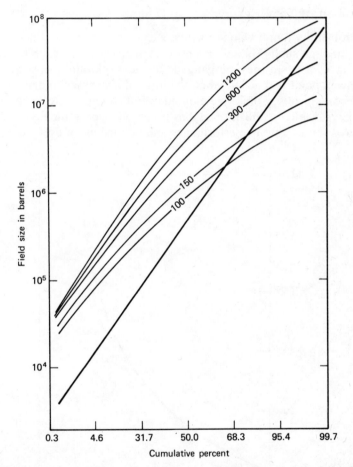

FIGURE 1.22. Hypothetical cumulative frequency distributions of sizes of 40th field discovered from populations of 100, 150, 300, 600, and 1200 fields. Heavy straight line represents cumulative frequency distributions of populations from which discovered fields have been drawn. Adapted from Kaufman and others (1975).

limited by these assumptions involved. For example, discovery probabilities could be set proportional to some fractional power of oil-field sizes instead of being directly proportional. Studies of the historical changes in the statistics of actual oil-field volumes and areas should be carried out in different regions so Monte Carlo simulation experiments could be conducted using assumptions known to accord with reality. Harbaugh and Bonham-Carter (1970) provide extensive detail in designing simulation experiments in geology.

Estimation of Total Reserves from Drilling Histories

The probability that a wildcat will be a dry hole has not yet been considered, although such probabilities provide a measure by which the relative maturity of a basin may be gauged. In the exploration of a petroleum province a point is reached when the ratio of successes to dry holes falls to an economically prohibitive minimum. For example, the discovery rate for wildcat wells in the Denver basin has changed radically with time. Both as a review of past performance and as a means of prediction, Haun

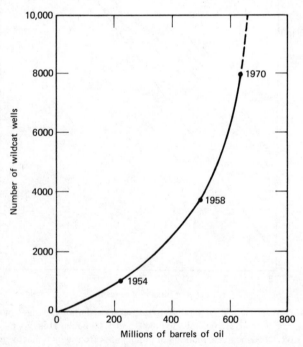

FIGURE 1.23.   Cumulative curve relating wildcat wells drilled versus volume of oil discovered in part of Denver basin. From Haun (1971).

(1971) plotted a cumulative curve (Figure 1.23) of number of wildcat wells drilled versus reserves discovered in the "*D*" and "*J*" sands in part of the Denver basin. Extrapolation of the curve to an asymptotic limit suggests that total reserves are about 660 million barrels which could be located by the drilling of 10,000 wildcat wells. The economic realities of wildcat drilling at any time may be measured by a tangent to the curve; the slope of the tangent line is the ratio between number of wildcat wells and the resulting volumes of discovered oil.

After a drilling history of 2000 wildcats the average volume of oil discovered in the basin was approximately 100,000 barrels per wildcat; the average yield fell to about 18,000 barrels per wildcat when 8000 wells had been drilled. These figures are not estimates of the probability of success or failure of an individual wildcat, a topic which is treated in a succeeding section. Rather, they demonstrate a useful measure of exploration efficiency, namely, the rate of change in the number of barrels discovered per exploratory well drilled in the basin.

In spite of the sharply decreasing efficiency of exploration over time, however, the distribution of volumes of oil fields discovered in the Denver basin remains steadfastly lognormal. A frequency distribution (Figure 1.11) of fields discovered through 1958 is nearly concordant with one based on discoveries up to 1971, except for the smaller fields. It is obvious, however, that the frequency distribution of oil-field volumes for fields that remain undiscovered at this time must differ drastically from that of the total population.

Relationship Between Area and Volume of Oil-Field Targets

Estimates of the distribution of oil-field volumes and cumulative reserves discovered may be related to numbers of wildcat wells drilled, providing valuable information for basin analyses. Such analyses attempt to estimate both the total volume of undiscovered reserves and the manner in which these reserves will be apportioned to fields of different sizes. However, an explorationist perceives undiscovered fields as targets whose areal shapes are to be located by wildcat drilling. Pool volumes are a measure of the magnitude of his success, if any. The relationship between field volume and field area is necessarily statistically positive, but is mathematically expressed by a fractional power rather than by a simple linear relationship. Areas are measured as the square of a linear dimension; volumes are a cubic function. Therefore, for fields where there is no systematic change in shape relative to size, the general equation can be written

$$V = kA^{3/2}$$

where V is volume, A is area and k is constant. The equation may be transformed and rewritten as follows:

$$\log V = \log k + 1.5 \log A$$

A plot of log transformed volumes of oil fields versus logs of areas may be expected to approximate a straight line. This characteristic is shown by log-log plots of oil fields in the Denver basin (Arps and Roberts, 1958; Haun, 1971) as in Figure 1.24. If there is a systematic change in shape with size, the linear relationship will still hold true, but the exponent will deviate from 1.5.

Geological Statistics of Oil and Gas Traps

Various agencies publish tabulations of oil reservoir data that include porosity, permeability, API gravity, and lithological information in addition to measures of field area and volume. Use of geological data provides a greater degree of diagnostic refinement in basin analyses. A petroliferous sedimentary basin can be considered as a stratified accumulation of lithic prisms, each having distinctive properties such as porosity, permeability, and other factors. As exploration proceeds within the basin, the regional variation in subsurface geology is more clearly defined and estimations may be made of the total volumes in horizons which have

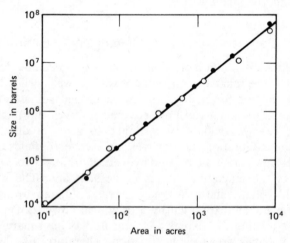

FIGURE 1.24.   Relationship between areas, and volumes of recoverable oil, in Denver basin oil fields. Open circles denote data from 338 fields obtained by Arps and Roberts in 1958, whereas solid circles denote data obtained by Haun from 557 fields in 1971. Adapted from Haun (1971).

favorable reservoir characteristics. By recording the geologic properties of oil fields as they are discovered, predictions may be made concerning the total volume of oil remaining in the basin keyed specifically to rock types in the various geologic horizons.

Mast (1970) tabulated properties of Illinois oil fields and summarized reserve information as related to geologic properties in a series of histograms (Figures 1.25 to 1.27). These distributions may be used for reserve estimation in Illinois. For example, about 60 percent of the total oil reservoir pore volume is contained in fine sandstone; 20 percent of such sandstones have permeabilities greater than 128 millidarcies (Figure 1.27).

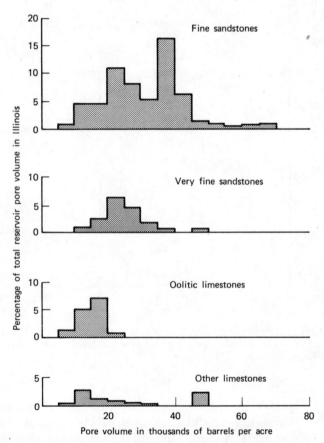

FIGURE 1.25. Histogram representing frequency distributions of pore volumes in thousands of barrels per acre in four lithologic types of oil reservoir rocks. Frequencies are expressed as percentages of total oil reservoir pore space in Illinois (histogram bars sum to 100). From Mast (1970).

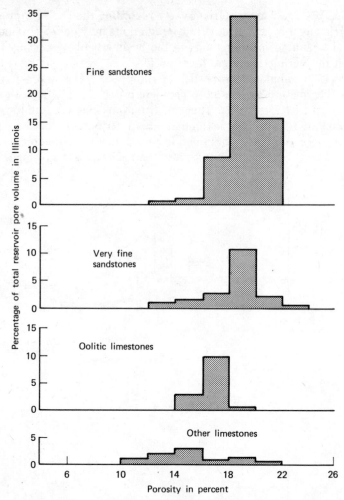

FIGURE 1.26. Frequency distributions of porosity by percentage in four lithologic types of oil reservoir rocks. Frequencies are expressed as percentages of total oil reservoir pore space in Illinois. From Mast (1970).

Similarly, about 45 percent of the reservoir sandstones (either fine or very fine grained) have more than 30 thousand barrels of reservoir pore volume per acre (Figure 1.25). A combination of these statistics with appropriate measures of the size and shape of reservoir formations enables a "budget accounting" to be made of the total reserves available in any basin. Naturally, the precision of such estimates is a function of both the infor-

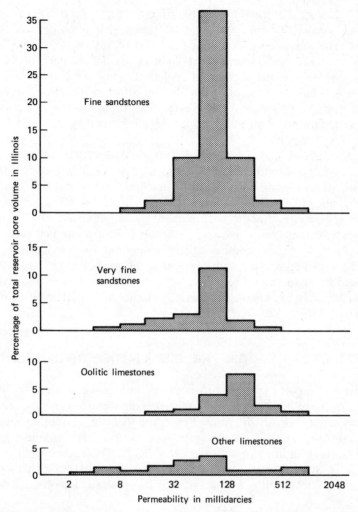

FIGURE 1.27. Frequency distributions of permeability in millidarcies in four lithologic types. Frequencies are expressed as percentages of total oil reservoir pore space in Illinois represented by each permeability class. From Mast (1970).

mation available and the validity of the geological model of the basin used in the estimate.

The frequency distributions shown in Figures 1.25 to 1.27 can be used for probabilistic forecasting. For example, the probability of a fine sandstone reservoir having a permeability greater than 128 millidarcies can be readily calculated. Inspection of the uppermost histogram in Figure 1.27 shows that about 60 percent of the total oil reservoir pore volume in Illinois consists of fine sandstone. This aggregate of 60 percent is subdivided into the various frequency classes of this histogram. Classes greater than 128 millidarcies (containing 10, 2, and ½ percent, respectively) represent about 12.5/60 or 20.8 percent of the total area of the histogram. Thus the overall probability of a fine sandstone reservoir in Illinois having a permeability of greater than 128 millidarcies is about 21 percent, and the probability of having a permeability less than 128 millidarcies is about 79 percent.

The histograms may be combined to yield a variety of probability estimates. For example, there is about a 55 percent probability that a sandstone (considering both fine and very fine sandstones) that forms an oil reservoir will have less than 30 thousand barrels of reservoir pore volume per acre, and about a 45 percent probability that it will contain more than 30 thousand barrels of reservoir pore volume per acre. This estimate is obtained by combining the histograms for fine and very fine sandstones in Figure 1.25 and computing the proportion of area of the combined histograms that represent reservoir pore volumes less than 30 thousand barrels, versus the proportion containing more than 30 thousand barrels per acre.

MODELS OF THE GEOGRAPHIC DISTRIBUTION OF OIL FIELDS

In order to pinpoint specific prospects as targets for wildcat drilling, a basic knowledge of the pattern of their areal distribution is necessary. Are oil fields geographically distributed in a random fashion, are they strongly clustered, or arranged in some systematic pattern? The simplest spatial model is that of a random distribution of fields. There are certain consequences of such a model that regulate the discovery of fields by the drilling of wildcat wells, if no account is taken of local geological conditions:

1. The probability that a wildcat will find a field is constant for any location in the area.
2. The probability of success in a drilling venture is independent of the outcome of previous wells.
3. The probability of the success of any individual well is fairly low, so discoveries can be considered as "rare events."

These rules specify that the frequencies of success in a series of small drilling programs will follow a Poisson distribution. The Poisson distribution describes the expected frequencies of one, two, three, or n objects per unit area or length, provided these objects are scattered randomly and are comparatively infrequent in occurrence.

A distribution of points generated from a random-number table is shown in Figure 1.28 and demonstrates that a random pattern of points may appear to be distinctly clustered when judged by casual observation. The common (and mistaken) conception of randomness is one of uniformity or comparative homogeneity, but close examination of a random collection of points shows an apparent mixture of all types of spatial arrangements, including local clusters and zones of seemingly aberrant dispersions. The casual impression of clustering is a reflection of the human tendency to look for patterns and to take special notice of "freak" clumps of points while ignoring nonclustered areas. By statistically comparing an observed frequency distribution with the corresponding Poisson frequencies predicted by a random model, an objective judgment may be made concerning whether oil fields are randomly dispersed.

If oil-field locations do not prove to be random, two simple alternative models may be proposed for the nonrandom pattern:

1. The fields show a distinct preference *against* clustering. Such a pattern might result from either a distinctive spatial ordering of individual fields, or some kind of "repulsion" mechanism mitigating against field congregation, though such a mechanism is difficult to conceive.

2. The fields exhibit pronounced clustering, due to localized developments of favorable trapping conditions caused by facies geometry, tectonic style, or hydrocarbon migration systems. The spatial pattern may be described by a negative binomial or other model of contagion.

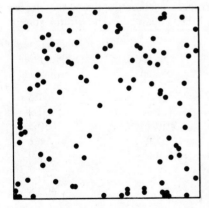

FIGURE 1.28. Random distribution of points generated from random number table.

Griffiths (1962) proposed that the negative binomial distribution is the most likely basic model for the distribution of natural resource targets (a mathematical exposition of the negative binomial distribution is provided by Kendall and Stuart, 1969, p. 130.) Griffiths (1966) studied the spatial arrangement of Kansas oil and gas fields as a check on his hypothesis. He "sampled" Kansas on a grid by selecting multiple sets of five wells with 5-mile spacing between wells. (The procedure is directly analogous to point counting in petrographic thin section analysis.) The number of producing oil or gas wells ("successes") in each set of five wells was accumulated as a frequency distribution. Theoretical frequency distributions were also computed assuming Poisson (random) and negative binomial (clustered) models. Griffiths' results, summarized in Table 1.5, clearly demonstrate the close fit of a negative binomial or clustered model to the observed data in contrast with alternative theoretical models.

The negative binomial is referred to as a "contagious" distribution and is used in biology and medicine to describe the dispersion of communities of organisms. When applied to the locations of oil fields, the model implies that individual fields tend to be clustered but the clusters themselves are randomly distributed. Clusters of fields may show gross alignments that are related to regional geological features. However, isolation of such trends requires other methods of analysis based on the conditional relationships between geologic variables and oil or gas accumulations.

Specification of a model for oil-field locations has a direct and obvious bearing on the choice of a search procedure to be used in exploratory drilling programs. The history of many petroleum provinces characteristically shows a distinctive pattern in the drilling of wildcats. The discovery of a new field is followed by a marked increase in drilling activity in the immediate vicinity, while drilling proceeds at a slower rate in rela-

TABLE 1.5.   Number of producing wells (successes) in sets of five wells selected by sampling across Kansas on a 5-mile grid. Data adapted from Griffiths (1966).

Number of Successes	Observed Frequency of Successes	Expected Poisson	Expected Negative Binomial
0	257	244.1	257.5
1	104	121.1	102.2
2	29	30.1	30.5
3	8	5.0	8.1
4	2	0.6	2.0
5	1	0.1	0.5
Total	401	401.0	400.8

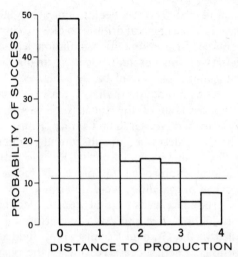

FIGURE 1.29. Frequency distribution relating probability of drilling successful wildcat well to distance from closest prior producing well in Graham County, Kansas. Horizontal line denotes mean of 11 percent.

tively virgin areas. This characteristic is good as an informal strategy because it implies a recognition of the clustered nature of fields. The prevailing view in industry is that there is a greater probability of finding a field adjacent to one already discovered than in currently unproductive areas.

This is expressed in the axiom of the petroleum industry that the best

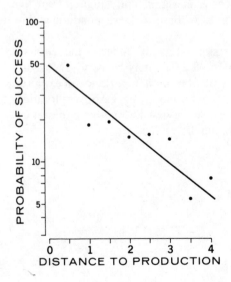

FIGURE 1.30. Frequency distribution employing same data as Figure 1.29, except that success probabilities are plotted on log scale instead of linear scale.

place to look for oil is where oil has been found. This leads to a drilling strategy that is based on leasing and drilling as close to initial discoveries as possible. Of course, in a competitive situation lease prices rapidly increase in the vicinity of a successful wildcat, so an operator who adopts this strategy must maintain a careful balance between lease costs and probability of success as a function of distance away from the success.

A probabilistic assessment of the success of this procedure as an exploration policy was made for Graham County, Kansas, the same area used for tests of methods described in subsequent chapters. All production comes from the Pennsylvanian Lansing-Kansas City Group; most reservoirs are structurally controlled. Intensive development began in the early 1950s. Discovery probabilities were determined by calculating the ratio of number of producing wells to total wells drilled, against distance to closest prior production. The analysis was performed on a file of all wildcat wells drilled in the Graham County area, divided into pre-1952 and post-1952 portions. Each successive well in the post-1952 file was examined in chronological order and its distance to the closest preexisting productive well calculated. The production status of the well was then determined and a count made in the proper distance category. The well was then placed in the preexisting well file and the process repeated on the next youngest well. The histogram in Figure 1.29 shows the probability of wildcat success as a function of distance from closest prior success. The probabilities remain roughly constant through time up to the present. The logarithmic plot in Figure 1.30 shows that the probability of success declines approximately exponentially with distance away from production. The probability of success is significantly higher than the average for the area if an operator can obtain a lease position within one-half mile of a discovery.

This simple probability analysis suggests that under some circumstances, a drilling policy of "follow-the-leader" may be very successful. However, such a procedure can only be of limited importance in exploration, as it obviously requires a leader who has a successful exploration policy of his own. In the next chapter, we review alternative exploration strategies which are based on probabilistic analyses.

CHAPTER 2
Exploration Strategies

SYSTEMATIC PATTERNS OF SEARCH

One approach to petroleum exploration does not consider the role of interpretative geology, but instead is concerned primarily with the mathematics of search theory. Exploration then becomes analogous to point counting a thin section which contains grains of a rare mineral. The likelihood that targets will be encountered (or fields will be discovered) is determined by their relative abundance and geometry (field sizes, shapes, and orientations), and by the geometry of the search pattern (spacing and arrangement of wells). Grid-drilling schemes are one form of search strategy. The theory of grid drilling has been extensively investigated by Griffiths and his associates (Drew and Griffiths, 1965; Griffiths and Drew, 1966; Drew, 1967; Singer and Wickman, 1969; Griffiths and Singer, 1971; Singer, 1972). Their investigations are based in part on the more general studies of search theory by Koopman (1956, 1957). In turn, search theory draws heavily on the principles of geometric probability (summarized by Kendall and Moran, 1963).

Most grid-drilling schemes approximate targets as simple shapes; for example, the geographic expanse of an individual oil field (or ore body) may be represented as an ellipse. If the long axis of the elliptical field is shorter than the spacing between successive wells in a drilling pattern, the field will be either missed entirely or hit only once. The probability of discovery is simply the area of the ellipse divided by the area of one cell of the grid. Singer (1972) gives a computer program for calculating these probabilities for various field dimensions and sampling patterns. Statistical tables have also been prepared for this purpose by Savinskii (1965) and by Singer and Wickman (1969).

In a grid-drilling project, a number of potential targets are assumed to exist in the exploration area. The problem then becomes one of defining the search pattern so that the maximum number of these targets are hit with the minimum number of exploration wells. If the targets are assumed to be uniformly distributed over the area of search, grid drilling is optimal in the sense that the greatest return will be achieved of any finite, patterned search effort. Similarly, if the explorationist has no knowledge of how the fields may be distributed through the exploration area, a grid-drilling scheme will allocate equal search efforts to all parts of the area. This again will be the optimal strategy, as compared to other patterned searches. These desirable properties have also been pointed out by Ellis and Blackwell (1959) and by Pachman (1966) in the context of geophysical exploration.

If the distribution of field sizes is known or can be inferred for an area, a grid-drilling density can be calculated that will maximize the yield or difference between the gross worth of oil discovered and the cost of discovery. Drew (1966) simulated the results of grid drilling in 15 major petroleum provinces of North America and found that the optimum drilling pattern had a spacing from 0.75 to 3.76 miles. The curve of yield versus spacing for the Texas Gulf Coast shown in Figure 2.1 is typical. Yields are relatively low if a short spacing is used between the lines of the drilling grid because of the inordinate number of wells that must be drilled. The return from additional discoveries of small fields found by the fine grid does not compensate for the higher drilling costs. Yields increase to a maximum with a 2-mile grid spacing, then decline with increasing space between wells because more fields are missed by the search.

Mickey and Jesperson (1954) and Singer (1975) have compared the efficiencies of various grid configurations. Comparisons include square versus rectangular grids, triangular or hexagonal grids, and diamond-shaped parallelograms.

Application of a grid-drilling program in a virgin area requires prior assumptions about the distribution of field sizes that will be encountered. Thus, grid drilling must rely on field-size statistics derived from studies of other, more mature areas judged to be geologically similar. In this regard, grid drilling is a natural extension of basin analysis. However, to date no major grid-drilling program has been undertaken for oil and gas, although similar programs have been used in the exploration for mineral deposits (Brown, 1962; Marshall, 1964; Imai and Itho, 1971; Lampietti and Marcus, 1971). The reluctance of explorationists to adopt grid-drilling schemes may reflect in part their unwillingness to abandon or ignore geologic information other than the statistics of oil-field sizes. Most explorationists believe (rightly or wrongly) that knowledge of the geology in

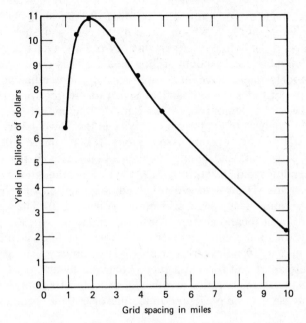

FIGURE 2.1.   Yield that would have accrued if Texas Gulf Coast had been drilled on grid spacing scheme. After Drew (1966).

an area will improve their success ratio. Perhaps more importantly, adoption of a grid-drilling approach would be difficult if not impossible in the intensely competitive, free-enterprise atmosphere that pervades the American oil industry. Still, grid-drilling concepts have merit and may be politically feasible for exploration over large areas under a single concession.

Random Drilling

Grid drilling is one of several possible exploration strategies. Random drilling is another. To our knowledge, no actual exploration programs have deliberately employed random drilling, although the cumulative exploration effort in some regions may approach randomness. A truly random strategy would involve some scheme for generating exploratory well locations at random, perhaps by drawing two geographic coordinate values from a random number table and drilling the exploratory hole at the location corresponding to the coordinate values. Such a strategy is almost totally foreign to existing exploration philosophy, and, of course, involves a total negation of the role of geology as a guide to exploration.

Nevertheless, an examination of a random drilling philosophy is warranted. One such study has been made by Menard and Sharman (1975), who contrasted the actual discovery history of large fields in the United States with hypothetical random drilling models. They concluded that the random models might have been appreciably more efficient than the industry's record of success, particularly in the search for giant fields.

Drew (1974) has employed a random drilling model as part of a "hindsight" analysis of exploration activity in the Powder River basin in Wyoming. The study encompassed about 27,000 square miles of the Powder River basin, and employed data obtained from the industry's efforts extending from 1889 through 1970, which resulted in the discovery of 154 oil fields. Drew's model involved the location of hypothetical exploratory wells at specified grid points (Figure 2.2). The grid points representing well locations were chosen at random, with the provision that no new location could be closer than some specified distance (which ranged from 2 to 10 miles) from a previous discovery. The grid points were ¼ mile apart and formed a rectangular meshwork.

Drew's simulations permit a number of probabilistic data to be calculated, and these may be presented in tabular or graphic form. For exam-

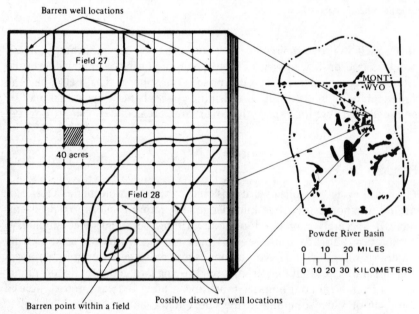

FIGURE 2.2. Diagram illustrating grid superimposed over Powder River basin. From Drew (1974).

ple, Figure 2.3 is a graph of probability values versus number of exploratory wells that pertain to the discovery of specific volumes of petroleum in the Powder River basin. These curves assume (*a*) that exploratory wells are located at random, subject to the constraint (*b*) that each new well is located not less than some minimum distance from previously drilled wells, and (*c*) that the exploration program begins in an unexplored or virgin basin. The probability curve for no discovery (gambler's ruin) is also shown.

The curves in Figure 2.3 could be useful in planning an exploration program in virgin basins in which the relationship between producible oil volumes and oil-field areas is believed similar to that in the Powder River basin, and in which the frequency distribution of oil fields is also believed similar. For example, about 100 exploratory wells would be needed to discover a total of 10 million barrels of oil with a probability of about 80 percent in such a basin.

Random drilling models are also useful for predicting the volume of oil extracted as a proportion of the total volume of oil present (i.e., the so-called total petroleum resource base, Q). Figure 2.4 contains curves for different total volumes of oil discovered (in millions of barrels) plotted according to number of exploratory wells drilled and different magnitudes assumed for the total petroleum resource base. A similar family of curves (Figure 2.5) may be developed relating the number of fields discovered versus number of exploratory wells drilled and total number of fields present in the resource base. A family of curves (Figure 2.6) representing the probabilities of no discovery (gambler's ruin) may also be calculated. The three families of curves in Figures 2.4 to 2.6 incorporate the assumptions of randomly located exploratory holes and an initially virgin basin whose oil-occurrence statistics are comparable to those of the Powder River basin.

Allais' Search Model

Allais (1957) was an early proponent of the use of statistical models for exploration and regional evaluation. He proposed that the Algerian Sahara be systematically explored for metallic deposits. His hypothetical exploration programs involve extrapolation of statistical data from maturely explored areas in other countries, assuming that the frequency distributions of metallic deposit occurrences could be reasonably applied in the Sahara. Allais divided his theoretical exploration programs into three stages. Stage 1 consists of an initial reconnaissance, which involves a "search for indicators." Stage 2 follows up the deposits detected in Stage 1 by test pitting and preliminary drilling. Stage 3 consists of the

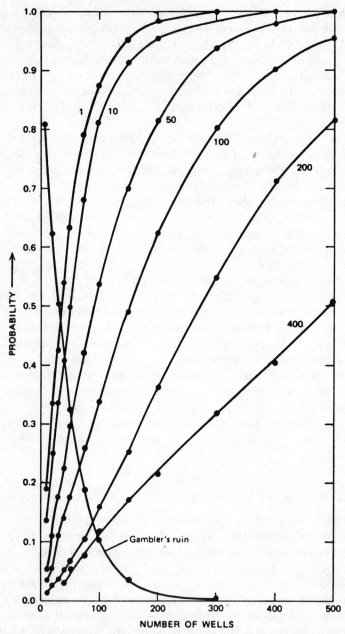

FIGURE 2.3. Probability curves for discovery of specific volumes of petroleum (in millions of barrels) via random exploratory drilling program in Powder River basin in its virgin state. From Drew (1974).

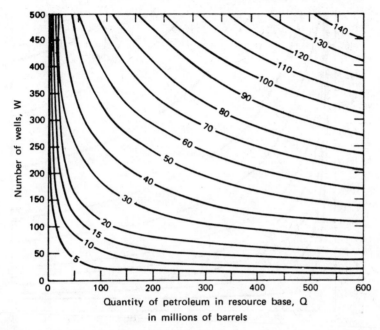

FIGURE 2.4. Family of curves representing aggregate volume of oil discovered with respect to number of randomly located exploratory wells drilled versus total quantity of oil initially present as resource base, Q, in basin whose oil-occurrence statistics are comparable to those of the Powder River basin. From Drew (1974).

actual preparation for mining, including detailed drilling and shaft sinking. A deposit is considered to be exploitable if the value of its annual production ranges between one million and one billion dollars. These two values define a workable range, the lower one reflecting the minimum effective size in a region such as the Sahara, and the upper limit representing the extreme above which a deposit would be unrealistically large. Allais' model involves the assumption that at each stage in the exploration the number of indicators detected or the number of deposits discovered follow a Poisson distribution (Kendall and Stuart, 1969, p. 125, provide a mathematical exposition of the Poisson distribution). The values of deposits discovered, however, are assumed to be lognormally distributed.

In assessing profitability of an exploration campaign on a regional basis, Allais employed three sets of estimated monetary values, the least favorable, the most favorable, and a median estimate that lies between the least and the most favorable. Allais also assumed that exploration procedures would be such that the largest deposit present has some reasonable probability of being discovered. He estimated the cost of a particular

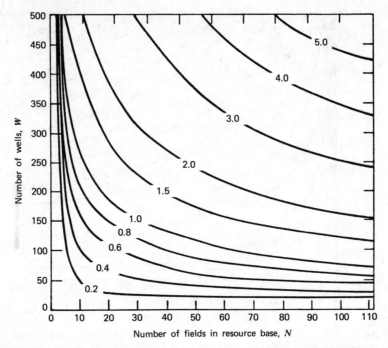

FIGURE 2.5.   Family of curves representing number of fields discovered with respect to number of randomly located exploratory wells drilled versus total number of fields present in resource base. From Drew (1974).

exploration campaign at $360 million. Combining this forecast of exploration costs and the assumed frequency distribution, Allais estimated that the probability of achieving a profit was only about one in three. Other levels of exploration effort can be assumed, however, and the probability of achieving a profit under those circumstances can be calculated.

Allais' conclusions applied to the Sahara are now out of date and his assumptions are subject to challenge, particularly concerning the frequency distributions of undiscovered ore deposits assumed to exist there. Other types of frequency distributions may need to be considered, a point emphasized by Slichter, Dixon, and Myer (1962), who compared observed and theoretical frequency distributions of ore deposits on an areal basis in parts of North America.

Allais' approach is of interest here, however, because regions that are relatively unexplored from a petroleum-search standpoint could be analyzed similarly. His hypothetical model has many similarities to the Engel model described next, although Allais' method is much less systematic and not concerned with the actual search pattern. Other economic

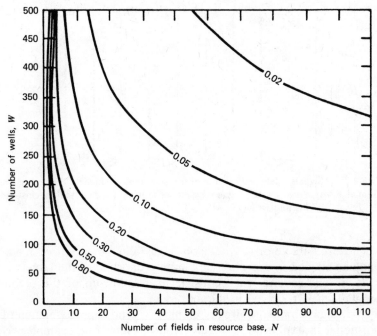

FIGURE 2.6. Family of curves for different probabilities of no discovery (gambler's ruin). From Drew (1974).

evaluation models that should be noted here have been proposed by Brinck (1967), Celasun (1964*a*, 1964*b*), de Guenin (1962), Kyrala (1964), Stachtchenko (1971), and Uhler and Bradley (1970).

The Engel Model

Various schemes for systematic search have been suggested; one of the more elaborate stems from the "Engel model" proposed by J. H. Engel (1957). Its application to a simulated exploration program for uranium was described by Griffiths and Singer (1971, 1973). The Engel model incorporates a two-stage search procedure when the targets are small relative to a large area over which they are assumed to be located at random. The Engel procedure involves a pattern of systematic traverses across the search area (Figure 2.7). During the search procedure, "detection equipment" receives "signals" from the targets; some of these "signals" are false and randomly distributed. In oil exploration the "signals" might consist of geophysical anomalies, oil seeps, or facies changes inferred from observations at outcrops. The true signals from real targets will tend

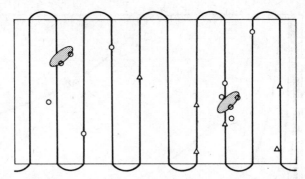

FIGURE 2.7. "Signals" detected in series of traverses in Engel model. "False" signals are represented by triangles, "true" signals by circles. Fields are detected by clustering of signals. From Griffiths and Drew (1964).

to be clustered above or near a target or oil field. An effective clustering technique will lead to detection of a large proportion of the targets in the first stage. The second stage consists of a follow-up of the seemingly favorable signals received in the first stage.

The petroleum industry already employs a crude approximation of the Engel model in exploring relatively virgin basins. For example, reconnaissance geophysical surveys are made initially; if favorable anomalies are found, detailed geophysical work is concentrated in the vicinity of the anomalies. Prospects that emerge from the follow-up or second-stage geophysical work may then be drilled. If a discovery is made, particularly one that appears to be large, additional drilling will cluster around the discovery. Thus industry practice leads to a clustering of discoveries and exploratory activity. Industrial behavior in the exploration of a new basin, however, tends to be far less systematic than the ideal of the Engel model.

The Engel model assumes that the targets are either randomly distributed or occur in randomly distributed clusters. Other assumptions are that the number of fields in the exploration region are known. This condition may be difficult to satisfy, although the number of fields may be estimated from cumulative frequency distributions of oil-field sizes or drilling success ratios as discussed earlier. Second, the probability that any given field will be detected by a first-stage search is a known constant. This constant may be approximated by superimposing a grid corresponding to the search pattern to be used over the map of a maturely developed basin, and by calculating the likelihood that one or more grid intersections will lie within the boundaries of the fields. A third assumption is that the first-stage detection of a field occurs within a known distance R of the actual location of the field. Therefore, a circle of radius R describes the area within which the field can be found. The final assumption is that the probability of a "false signal" is constant per unit of area over the region being explored and that it follows a Poisson distribution. The frequency distribution of combined true and false signals is the conditional probability of the combined Poisson and binomial distributions (the mathematics of these distributions in combination are treated by Kendall and Stuart, 1969, pp. 128 and 146.)

If economic assumptions are appended to the Engel model, it is possible to compute the profit or loss for a given search strategy such as a particular grid-drilling program (Griffiths, 1966). Economic assumptions include drilling costs, the value of the fields to be discovered (perhaps expressed as discounted net cash flows), and the lease or land acquisition costs.

UNIT REGIONAL VALUE ANALYSIS

Griffiths (1969) has suggested that unit regional statistics are useful guides in estimating the production potential of the relatively unexplored parts of a partially explored region. Unit values may be expressed in terms of the value of a unit of the earth's crust, in dollars per square mile or dollars per cubic mile. Griffiths has computed the unit regional value of extractable commodities (coal, petroleum, metals, and sand and gravel) per square mile for all of the 50 states of the United States. The value of all mineral commodities produced in the United States between 1880 and 1964 is approximately $400 trillion. Considering that the area of the United States is slightly more than 3 million square miles, on the average each square mile yielded about $132,309 during this period. Of this average, fuels accounted for $86,572 per square mile, nonmetallics $24,632 per square mile, and metals $21,105 per square mile.

These averages for the United States as a whole, however, fail to convey the enormous range in productivity from state to state, or for that matter, from area to area within individual states. Figure 2.8 is a diagram in which states are ranked according to their unit regional value. Note that the diagram is also a frequency distribution in which logarithms (to base 10) of the unit regional values per state are plotted. The distribution is approximately lognormal, a feature more clearly emphasized when the distribution is plotted in cumulative form (Figure 2.9).

A single unit regional value for an entire state is of value in exploration planning. For example, Alaska's extremely low unit value of only $3,483 per square mile for mineral commodities produced from 1911 through 1964 suggests that Alaska was grossly underexploited during that period

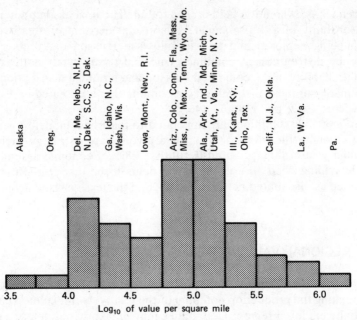

FIGURE 2.8. Frequency distribution of mineral production per square mile by states. Adapted from Griffiths (1969).

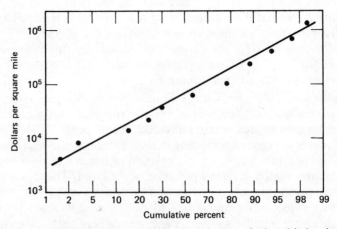

FIGURE 2.9. Cumulative frequency distribution of value of mineral industries products from 1911 to 1964 by states for United States expressed in dollars per square mile. Points plotted on graph represent midpoints of 11 histogram bars of Figure 2.8. Adapted from Griffiths (1969).

by comparison with the rest of the United States. Events since 1964, such as massive oil discoveries on Alaska's North Slope, confirm this indication of underexploitation prior to 1964.

The averages for entire states are of little use in guiding exploration within a state. Unit regional values can be computed, however, over a state on a geographic cell-by-cell basis. In Figure 2.10, Kansas and parts of adjoining states have been divided into a grid of cells. Each point on the grid roughly coincides with the center of a county, as production statistics are compiled on a county-by-county basis. This introduces some slight distortions to accommodate the geography of the counties, and the ideal of one grid point per county was not possible everywhere in the region.

Variations in production of extractable mineral commodities over the region are shown by a contour map of the unit regional values, or the logarithms of the unit regional values (Figure 2.11). The log transformation has a subduing effect on the extreme values, and the resulting data are more readily contoured.

An alternative approach in mapping unit regional value is to estimate the values at a succession of points over an area. The estimation involves the use of multiple linear regression. For example, the value of oil and gas produced can be regressed upon a variety of geologic factors. Furthermore, nongeologic factors can be incorporated in the regression equation.

An example is provided by Singer's study (1971) of oil and gas production in California, in which the base-10 logarithm of the value of oil and gas produced in 1968 per square mile was regressed on eight geologic variables and two economic variables, as listed in Table 2.1.

The geologic factors (except for average Bouguer gravity) were estimated for each California county by analyzing the proportion of outcrops on the published geologic map of the state.

Singer's regression equation "explains" 52.7 percent of the variation in oil and gas production. Conversely, 47.3 percent remains unexplained. Although generalized, the geological variables in combination with the two economic terms succeed in representing slightly more than half the variation in hydrocarbon production by counties.

The regression estimates of oil and gas production can be contoured. The resulting map (Figure 2.12) is a generalized portrayal of the geographic variation in hydrocarbon worth as predicted by geologic factors. The highly productive region of Los Angeles and Ventura counties (enclosed by contour 4), for example, stands out. If the dollar values were contoured rather than their logarithms, such regional variations would be much more pronounced.

Because only slightly more than half of the variation is accounted for by Singer's regression equation, there is a marked difference between the

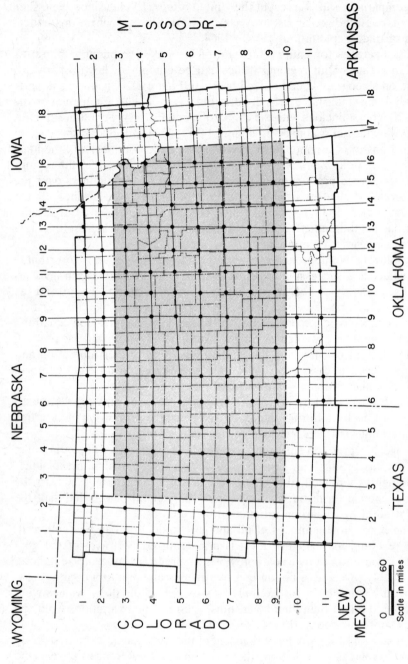

FIGURE 2.10. Orthogonal grid points fitted to Kansas and two tiers of counties in adjoining states for purposes of computing unit regional values. From Griffiths (1969).

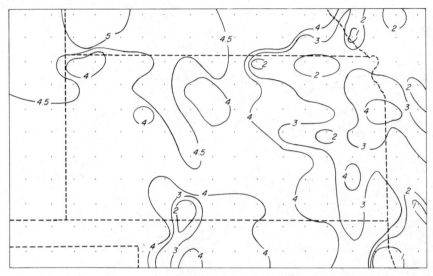

FIGURE 2.11. Contoured surface of logarithms (base 10) of dollar value per square mile of Kansas and adjacent states based on value of extractable commodities produced through 1960. From Griffiths (1969).

TABLE 2.1. Regression equation relating value of oil and gas production per square mile to eight geologic variables and two economic variables. From Singer (1971).

$$Y = -3.0746 + 0.0007925X_{11} + 0.9740X_{12} + 0.05233X_{13} + 0.02688X_{15}$$
$$+ 0.04462X_{16} + 0.6388X_{17} + 0.1460X_{18} + 0.005243X_{19} + 0.003471X_{21}$$
$$- 0.03671X_{23}$$

where

Y = $\log_{10}$ value of petroleum and gas production per square mile (1968 dollars)

X_{11} = Gross National Product (1968 dollars)

X_{12} = $\log_{10}$ population per square mile

X_{13} = percentage of Cenozoic marine sediments

X_{15} = percentage of Cenozoic nonmarine sediments

X_{16} = percentage of Mesozoic marine sediments

X_{17} = percentage of Paleozoic marine sediments

X_{18} = percentage of pre-Cretaceous volcanic and metavolcanic rocks

X_{19} = percentage of Mesozoic "granite" × percentage Mesozoic marine rocks

X_{21} = percentage of Mesozoic "granite" × percentage pre-Cretaceous volcanic and metavolcanic rocks

X_{23} = gravity (average Bouguer anomaly)

FIGURE 2.12.   Generalized contour map of California showing base-10 logarithms of dollar values (in 1968 dollars) of oil and gas production per square mile predicted by regression equation. Adapted from Singer (1971).

value predicted for each county and that actually recorded. The differences are termed *residuals,* and may be plotted and mapped much in the same manner as trend-surface residuals described in succeeding chapters in this book. Figure 2.13 is a generalized contour map of residuals from Singer's regression equation. Our interest here is focused on the residual lows, or negative values, which include localities that may be regarded as "underproductive" from a statistical standpoint and which may warrant additional exploration because their predicted potential has not been realized.

The number of geological variables in Singer's study is quite limited, but the method does provide an overview based on surface geology. It is of interest as an example of a technique that might be employed in other regions, particularly where some information is available on the surface geology but there is little or no subsurface geological information.

Sinclair and Woodsworth (1970) have employed multiple linear regression in a similar manner for evaluation of metal mining exploration in

FIGURE 2.13. Generalized contour map of California showing residuals represented as base-10 logarithms of value of 1968 oil and gas production per square mile in 1968 dollars. Adapted from Singer (1971).

British Columbia. In their study they regressed metal production, given in dollars, on geological variables measured on geologic maps in a series of maturely explored "training" cells. The regression equation was then used to forecast the metal potential in "target" cells located in areas that have undergone relatively little exploration.

USE OF "TRAINING" AND "TARGET" AREAS IN REGIONAL VALUE MAPPING

The technique of mapping unit regional value may be employed in an exploration context by statistical comparison of "training" and "target" areas. This concept in a petroleum exploration context is developed in detail in Chapters 5 to 7. It is useful here, however, to review an application by Harris (1969*a*) to exploration for metals in parts of Alaska. Harris had previously compiled data on the distribution of metals in two large regions (Figure 2.14) embracing parts of Utah, Arizona, New Mexico, and

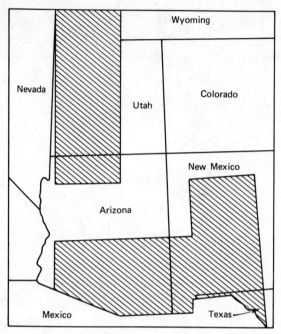

FIGURE 2.14. "Training" areas in southwestern United States used by Harris (1968) for statistical forecasts of metal occurrence in Alaska.

Texas, which serve as training areas. By comparison with Alaska, the two training areas represent relatively well-endowed and well-explored regions. Alaska, by contrast, is much less well explored. Whether the mineral wealth in Alaska is equivalent to the training area is a major question. If the distribution of metals in the two areas is equivalent, then the training areas can be used to statistically forecast Alaska's undiscovered metal resources.

Harris divided the training areas and parts of Alaska into square cells, each 20 miles on a side. The dollar worth of metal production was determined for these cells from records for mining districts in each cell. Metal prices prevailing in 1964 were used to transform all production figures to dollars. The dollar figures per cell were tabulated as frequencies (Table 2.2), permitting the cell-value frequencies in the training areas to be compared with those of Alaska. Table 2.3 provides a comparison between frequencies in monetary classes in Alaska, in contrast to the expected frequencies. The most conspicuous feature of Table 2.3 is the surplus of cells of value class 1 ($0 to $10,000), and the marked deficiency of cells in all higher classes except value class 6. This comparison

TABLE 2.2.   Frequency data in terms of dollars per cell in which data from cells in training areas in Utah, Arizona, and New Mexico are compared with cells in Alaska. From Harris (1969a).

		Frequencies	
Value Class	Value ($)	Utah, Arizona, and New Mexico (Training Areas)	Alaska (Target Area)
1	0 to 10^4	265	460
2	10^4 to 10^5	24	0
3	10^5 to 10^6	30	7
4	10^6 to 10^7	30	12
5	10^7 to 10^8	22	8
6	10^8 to 10^9	6	6
7	10^9 to 10^{10}	9	0
8	10^{10} +	1	0
Totals		387	493

suggests great rewards for exploration for metals in Alaska. The expense of exploration in Alaska posed by the difficulties of access and attendant higher costs, however, may dampen the enthusiasm for exploration there under present economic conditions.

The comparisons presented in Table 2.2 and 2.3 are based solely on metal production statistics transformed into dollars. No geological con-

TABLE 2.3.   Expected versus known occurrences of value classes in 493 cells in Alaska. Location of cells is shown in Figure 2.15. Relative frequencies listed in the second column are based on data from the training areas in southwestern United States. From Harris (1969a).

Value Class	Relative Frequency	Expected Occurrence in Alaska (493 Cells)	Known Occurrence	Difference
1	.685	338	460	122
2	.062	31	0	−31
3	.078	38	7	−31
4	.078	38	12	−26
5	.057	28	8	−20
6	.015	7	6	−1
7	.022	11	0	−11
8	.003	2	0	−2

siderations are incorporated. Harris has shown, however, that it is also possible to statistically consider a variety of geological details on a regional basis by systematically extracting data from conventional geologic maps which have been divided into geographic cells of uniform size (20 × 20 miles square). These geological variables are listed in Table 2.4 and are similar to those employed by Singer, described previously.

The relationship between these geological variables and metal production was calculated by computing a discriminant function. The

TABLE 2.4. Geological variables employed by Harris (1969a) in statistical comparison between training areas in southwestern United States and parts of Alaska.

X_1 = Percentage of cell area consisting of sedimentary rock.

X_2 = Percentage of cell area consisting of Precambrian-Paleozoic igneous intrusives.

X_3 = Percentage of cell area consisting of Mesozoic-Cenozoic igneous intrusives.

X_4 = Percentage of cell area consisting of igneous extrusive rocks.

X_5 = Percentage of cell area consisting of metamorphic rocks.

X_6 = Number of high-angle faults 8 miles or less in length.

X_7 = Number of high-angle faults greater than 8 miles in length.

X_8 = Number of low-angle faults 8 miles or less in length.

X_9 = Number of low-angle faults greater than 8 miles in length.

X_{10} = Number of anticlines 8 miles or less in length.

X_{11} = Number of anticlines greater than 8 miles in length.

X_{12} = Number of high-angle fault intersections.

X_{13} = Number of low-angle fault intersections.

X_{14} = Length of contact of undifferentiated igneous intrusives with sedimentary rocks.

X_{15} = Number of exposures of the above contact.

X_{16} = Length of contact of igneous intrusives with igneous intrusives.

X_{17} = Number of exposures of the above contact.

X_{18} = Length of contact of Mesozoic-Cenozoic igneous intrusives with Precambrian-Paleozoic igneous intrusives.

X_{19} = Number of exposures of the above contact.

X_{20} = Length of contact of Mesozoic-Cenozoic igneous intrusives with metamorphic rocks.

X_{21} = Number of exposures of the above contact.

X_{22} = Length of contact of Precambrian-Paleozoic igneous intrusives with metamorphic rocks.

X_{23} = Number of exposures of the above contact.

X_{24} = Number of igneous dikes.

X_{25} = Percentage of cell area covered by Quaternary alluvium, water, or ice.

coefficients of the discriminant function were calculated on the basis of "experience" in the training areas (Figure 2.14) in southwestern United States. Given these coefficients, a forecast of metal production in each of the cells in Alaska was made, as the same geological variables in Alaska were also measured on a cell-by-cell basis. The use of discriminant function analysis is described in Chapter 6, and we delay any explanation of the method until then. In Harris' application, the method permits the probability of correctly assigning a cell to specified value classes to be calculated. Figure 2.15 shows the probabilities of correct assignments to value class 3 ($100,000 to $1,000,000). Probability of the correct assignment to other value classes can be calculated similarly.

Related papers by Harris that deal with the use of regression and discriminant methods for statistical forecasting of mineral potential were published in 1964, 1965a, 1965b, 1968, and 1969b. A report by Harris and Chrow (1969) also describes a probabilistic regional appraisal of Alaska's base and precious metal resources.

It may be inappropriate to use surface geology as an exploration guide for hydrocarbons, as Harris has done for metals. However, the statistical methodology is definitely appropriate for the appraisal of sedimentary basins early in their exploration history. In petroleum exploration there is the potential of an even more useful comparison which could be made between a training and an exploration area if geophysical or remote sensing data were available. The mature training area would provide information (or "ground truth") about the relation between response of the remote sensors or geophysical measurements and the presence or absence of oil as revealed by drilling. A discriminant function could be calculated which would be the combination of remote sensing or geophysical responses that best divided the training area into its known dry and productive parts. If this discriminant function were then used in the exploration area, where perhaps only remote sensing observations had been taken, predictions could be made of the likelihood of success of future drilling. With satellite reconnaissance, such a methodology could be used not only for prospecting in remote areas, but also for evaluating the resources of competing groups or even nations. The success with which this could be done would depend on the degree of similarity between the training and exploration areas and the relative effectiveness of the measurements used.

Use of Probability Index Maps

Probability index maps have been used in exploration for minerals, and in principle could be applied to petroleum exploration. The method also

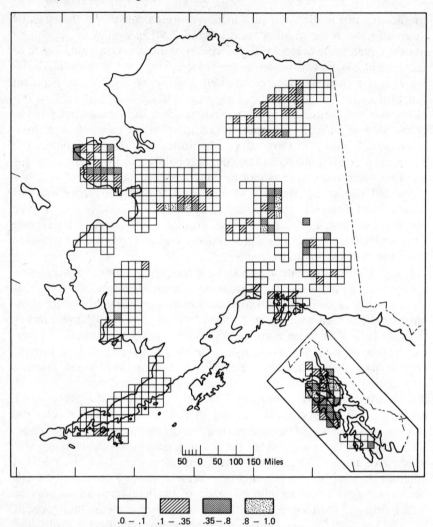

FIGURE 2.15.   Distribution of geographic cells in Alaska for which probability of belonging to value class 3 has been calculated. From Harris (1969*a*).

involves use of training areas that are statistically analyzed to provide a background for extension to target areas. In mining applications, a grid of cells of equal size is superimposed over the region to be analyzed. Some of the cells, which contain ore deposits of known size and grade, are selected as control or training cells. The geology of the region is assessed statistically on a cell-by-cell basis and the degree of geological similarity between training and target cells is used as an index of the probability of occurrence of ore deposits. Results can be displayed as contour maps of the probability index values.

An example is provided in a study of the copper potential in the Abitibi area of Ontario and Quebec by Agterberg and his colleagues (1972). They divided the Abitibi area into a hierarchy of large cells (40 × 40 kilometers), each of which in turn was divided into 16 smaller (10 × 10 kilometers) cells. Probability indexes were computed for each of the large cells. If a large cell has a probability index of 2, for example, this indicates that two of the smaller cells within it are forecast to contain at least one copper deposit containing 1000 tons of copper or more. Thus the probability index expresses the minimum number of copper deposits expected in the large cell. In turn, the copper potential can be expressed as a tonnage forecast by multiplying the probability index by the average tonnage of copper present in the 10 × 10 kilometers training cells.

Agterberg's calculation of the probability index involved tabulation of geological variables that were systematically measured in the cells of the Abitibi area, much as both Singer and Harris have done. Agterberg's variables include the proportion of outcrops of seven major rock types, which include granitic rocks, mafic intrusions, ultramafics, acid volcanics, mafic volcanics, Archean sedimentary rocks, and metamorphosed sedimentary rocks. Proportions were determined by point counting a geologic map at a scale of 1:250,000. In addition to rock proportions, the total length of layered iron formation per cell, average Bouguer gravity anomaly for the cell, and the regional aeromagnetic anomaly at the center of each cell were used as variables.

Probability index values are calculated by multiple linear regression, in which copper tonnage estimates in the training cells are regressed on the 10 geological variables. The coefficients of the regression equation were then used to estimate the copper potential as a probability index value in the target cells. Because the probability index contours are forecasts, they are sensitive both to the geological parameters chosen and to the choice of training cells. For example, Figure 2.16 shows three different probability index maps of the Abitibi area, each based on different sets of training cells.

The use of probability index contour maps is of interest because the technique could be adapted for oil exploration. For example, relationships shown over a region by a succession of maps, such as structure contour and lithofacies maps, could be used to calculate probability indexes based on statistical relationships in training areas containing known oil fields.

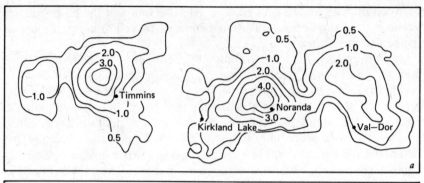

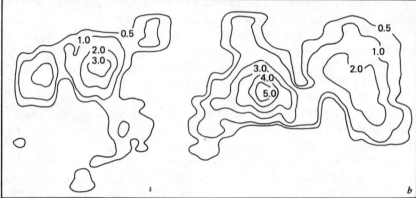

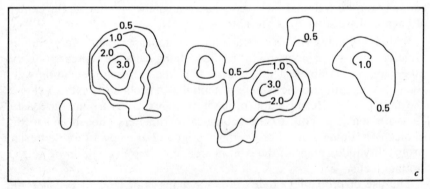

FIGURE 2.16. Probability index maps of copper potential in the Abitibi area, Ontario, and Quebec. Expanse of map is 170 × 590 kilometers. From Agterberg and others (1972). (*a*) Map based on 50 training cells. (*b*) Map based on 19 training cells near Noranda, Quebec. (*c*) Map based on 8 training cells near Timmons, Ontario.

Subjective Probability in Regional Analysis

Subjective probability estimates have been used by Ellis, Harris, and VanWie (1975) to estimate uranium resources on a regional basis in New Mexico. Their estimate involved a survey of forecasts by 36 geologists, in terms of informed guesses or subjective probabilities assigned by them to number of deposits, ore tonnage, and ore grade. The geologists' estimates were weighted by a self-appraisal index based on their knowledge of uranium deposits in the region and then combined in various ways. Detailed estimates were made by dividing the state into 62 geographic cells. The richest deposits are forecast in the northwestern part of the state. The aggregation of eight cells encompassing the San Juan basin is estimated to contain 92 percent of the undiscovered uranium in the entire state. In addition, estimates of the depth to possible ore deposits and their thickness allowed other estimates to be made of the costs of developing, extracting, and milling the ore. It is forecast that 80 percent of the U_3O_8 to be found in New Mexico will be available at a cost less than $15 per pound (in 1974 dollars), and about 98 percent at less than $30 per pound. Undiscovered uranium resources in New Mexico are estimated at over 200 million tons of material, and to contain almost 500,000 tons of U_3O_8. The study of Ellis, Harris, and VanWie is significant from a petroleum-exploration viewpoint because it is an example of the use of subjective probabilities to yield a regional resource appraisal of mineral occurrences in sedimentary rocks.

CONDITIONAL ANALYSIS

In Chapter 1, we emphasized that much of this book is devoted to "conditional analysis." As used here, conditional analysis refers to the calculation of probabilities that are conditional on particular geological features or properties. For example, in most oil-producing areas, the probabilities of discovering oil conditional on the presence of an anticline would differ from probabilities that are conditional on the presence of some other structure.

A second point stressed earlier is that estimates of oil-exploration probabilities should be based on frequencies of occurrence. Thus, if probabilities for an oil discovery are to be conditional on a certain type of geologic feature, we must amass sufficient information to ensure that frequencies of different oil-discovery outcomes can be tabulated with reference to the particular geologic feature. These frequencies can then be used as estimates of the conditional probabilities. The process is simply a systematic formalization of our collective past experience.

One of the major objectives of this approach is to estimate the outcome probabilities attached to specific oil prospects. In using past experience as a guide to the appraisal of prospects to be drilled in the future, comparisons must be made with past prospects as they appeared *before* they were drilled. A geologic interpretation made after drilling is not appropriate, because hindsight knowledge (or rationalization) is substantially better than our knowledge of a prospect before it was drilled. Also, exploration technology has changed progressively so the efficiency of techniques such as geophysical well logging and seismic surveying are vastly improved today. In spite of these technical advances, some consistency in interpretation must be present to use past results to forecast future outcomes.

Unfortunately, the prevailing practices in the oil industry have not involved the systematic collection of interpretative geologic data on prospects. When a prospect is being generated, the relevant information is highly proprietary and is restricted in its circulation. After a prospect is drilled, the previous interpretation is "corrected" and the initial interpretation or interpretations are generally lost from the record. Within a typical company, there seems to be little awareness of the value of these early interpretations, and there is certainly no effort to make such records available to the industry as a whole.

Early KOX Experiments in Conditional Analysis

The Kansas Geological Survey undertook a series of experiments that involved the cross-tabulation of wildcat well outcomes with subsurface geological features. The work was part of the KOX (Kansas Oil Exploration) project, much of which is described later in this book. The initial experiments were carried out by Alfredo Prelat (1974) in an area in Stafford County, Kansas (Figure 2.17). The objective of Prelat's study was to develop a method for quantitatively estimating the outcome probabilities for wildcat wells, based on visual interpretation of structure contour maps. The probability estimates which he obtained reflect the combined probabilistic influences of both perceived geology and oil occurrence. Prelat's method may be described as a "reexperience" approach, in which the geology of the area has been repeatedly reinterpreted, incorporating progressively more well data into the analysis in the same sequence in which the data actually became available. Prelat, in effect, has recreated the exploration history of the study area through 40 years of drilling development, from 1930 to 1970.

Stafford County is well suited for such a postmortem analysis. The area studied is a square, four townships on a side (24 × 24 miles). More than 5000 wells have been drilled in the area since exploration began in 1930,

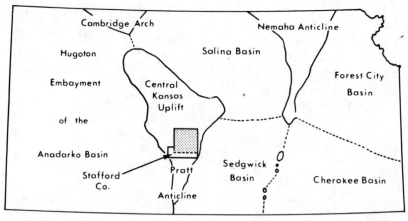

FIGURE 2.17. Index map showing location of KOX experimental area in Stafford County, Kansas. Map also shows boundaries of tectonic provinces. From Prelat (1974).

and many oil fields (Figure 2.18) have been discovered. Of the available wells, a total of 1326 that were drilled before the end of 1970 were used in the KOX project (Figure 2.19). These 1326 wells include 765 wildcats, in addition to 561 field wells selected to provide an equable geographic distribution of data points. The number of available wells, of course, has increased with time.

Since Prelat's study involved manual interpretation of contour maps, it was necessary for him to classify structures as they were visually perceived before they were drilled so that their drilling outcomes would be tabulated with respect to structure. The comparison is thus one of *predrill* structure with *postdrill* results. The postdrill outcomes are expressed as dry holes or as field discoveries of various magnitudes.

Prelat used a hierarchical classification with broad features defined as structural highs or structural lows, and subsidiary features as anticlines, noses, homoclines, and synclines. The classification thus involved dividing structures into two categories and then subdividing these categories again, as illustrated in Figure 2.20. This classification is flexible, but because of its branching nature the number of outcomes in the finer categories may be too small for reliable estimation of outcome probabilities. However, finer categories can be combined into the broader categories if necessary.

The outcomes of wildcat drilling were classified into five classes, as dry or one of four magnitudes of oil-field discovery. Production magnitudes were defined as the number of barrels of oil produced from a field in the initial five years following its discovery, regardless of the number of

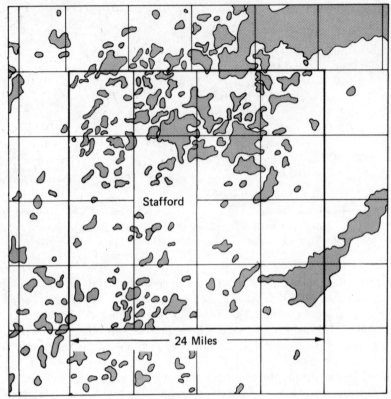

FIGURE 2.18. Stafford County, Kansas, showing oil fields discovered by end of 1966. KOX experiments were conducted in area outlined by heavy lines. From Prelat (1974).

field wells drilled. This definition allows comparison of fields discovered at different times, such as those found early in the area's history with those discovered more recently. The outcome categories consist of a progression of sizes based on powers of ten: less than 10^3 barrels (dry or essentially dry), 10^3 to 10^4 barrels (very small), 10^4 to 10^5 barrels (small), 10^5 to 10^6 barrels (medium), and greater than 10^6 barrels (large). Table 2.5 is a cross-tabulation of the four structural categories with five production outcome classes.

Example Calculation of Well Outcome Probabilities

Figure 2.21 is a structure map of the top of the Lansing Group, contoured by computer using data from all wildcat and selected field wells drilled prior to the end of 1940. Geologic interpretations based on this map were

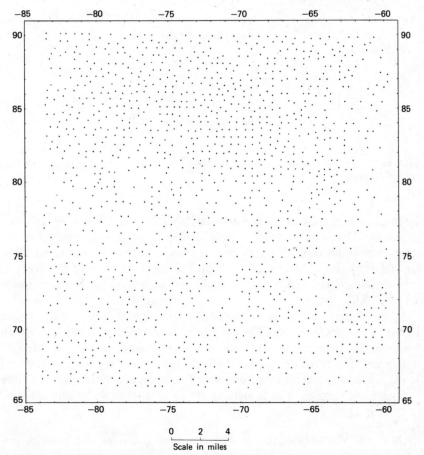

FIGURE 2.19. Location of wells drilled before end of 1970 in Stafford County area which were used in KOX study. From Prelat (1974).

cross-tabulated against wildcat well outcomes during the period January 1, 1941 to December 31, 1945. The wells shown in Figure 2.21 are those exploratory wells drilled during this period (*postdating* the wells on which the structural map is based). The number associated with each successful wildcat (solid circles) indicates the number of barrels of oil produced during the first five years from the field discovered by that well. Table 2.6 is an analysis of this map and shows outcomes arranged by columns. The estimates of outcome probabilities in part *B* of Table 2.6 are obtained by dividing the frequencies by the row totals in part *A* of the table. For example, the frequency of dry holes on anticlines during this

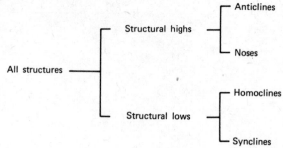

FIGURE 2.20. Hierarchical scheme used by Prelat (1974) to classify structures in Stafford County, Kansas.

interval as recorded in the top row of Table 2.6A is 2/4 = 0.50. In other words, two out of four wildcat wells drilled on anticlines resulted in dry holes during this time. From the bottom row (all structures), the probability of a dry hole is 114/133 = 0.86, whereas the probability of discovery of a field yielding 10^3 to 10^4 barrels in the initial five years of production is 1/133 = 0.01, and so on.

Many of these probability estimates are not reliable because they are based on insufficient data. The calculated probability of a dry hole on an anticline, for example, is based on only four exploratory wells located on anticlines. The broader structural classes (structural highs and structural lows) include more observations and consequently yield more reliable probability estimates. Fields yielding 10^5 to 10^6 barrels on an anticline and

TABLE 2.5. Cross-tabulation of predrill structural classes with postdrill outcomes in Stafford County experimental area. Outcome categories are dry or consist of oil-field production classes based on barrels of oil produced at each field during the initial five years following discovery by a wildcat well. From Prelat (1974).

Exploratory Well Outcomes (Initial Five-year Cumulative Production) (barrels)	All Structures			
	Structural Lows		Structural Highs	
	Syncline	Homocline	Nose	Anticline
Dry: 0 to 1000				
1000 to 10,000				
10,000 to 100,000				
100,000 to 1,000,000				
Greater than 1,000,000				

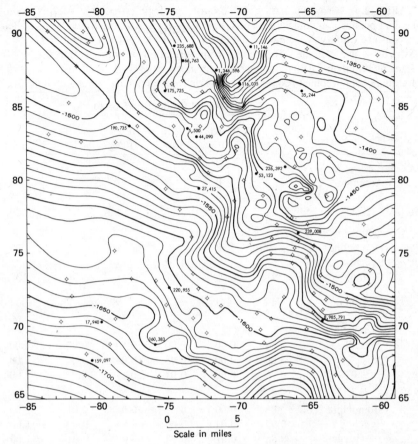

FIGURE 2.21.   Map of top of Lansing-Kansas City Group in Stafford County experimental area, based on well data available before end of 1940. Locations of wildcat wells completed from January 1, 1941 to December 31, 1945 are shown as dry holes (open circles) or discoveries (solid circles). Number of barrels of oil produced during first five years from each field is indicated next to its discovery well. From Prelat (1974).

on a nose involve frequencies of 2/4 and 2/25, respectively. When these two categories are combined as the category, structural highs, the frequency is 4/29, which is a more reliable estimate of the probability.

In practice, a structural contour map usually is updated and recontoured immediately after receipt of new information rather than every five years. However, the same procedures can be followed to produce frequencies of success, and in turn, probability estimates.

A difficulty in Prelat's approach is the consistent classification of the

TABLE 2.6. Cross-tabulation of predrill geologic structure interpretations against postdrill outcomes in barrels of oil discovered for exploratory wells drilled in Stafford County experimental area from January 1, 1941 to December 31, 1945. Structures are classified according to interpretation of contour map on top of Lansing based on data available as of December 31, 1940. *A*. Well frequencies. *B*. Estimates of outcome probabilities based on frequency data. After Prelat (1974).

	Postdrill Production Outcomes					
	Dry: 0 to 10^3	10^3 to 10^4	10^4 to 10^5	10^5 to 10^6	$> 10^6$	Totals
A. Structural highs						
Anticline	2	0	0	2	0	4
Nose	20	0	2	2	1	25
Total	22	0	2	4	1	29
Structural lows						
Homocline	60	0	5	5	1	71
Syncline	32	1	0	0	0	33
Total	92	1	5	5	1	104
Grand total	114	1	7	9	2	133
B. Structural highs						
Anticline	0.50	0.00	0.00	0.50	0.00	1.00
Nose	0.80	0.00	0.08	0.08	0.04	1.00
Aggregate	0.76	0.00	0.07	0.14	0.03	1.00
Structural lows						
Homocline	0.85	0.00	0.07	0.07	0.01	1.00
Syncline	0.97	0.03	0.00	0.00	0.00	1.00
Aggregate	0.88	0.01	0.05	0.05	0.01	1.00
All structures	0.86	0.01	0.05	0.07	0.02	1.00

continuous range of geologic structures in terms of distinct categories. When does an anticline become a nose, and when does a nose become a homocline? This uncertainty contributes an additional probabilistic influence in the analysis. A partial remedy employed by Prelat was to enclose each wildcat location with a geographic cell of arbitrary size (Figure 2.22). The size was dependent on the perceived complexity of the structure, reflecting different amounts of information available at different times. The structure seen within the area around each well was assigned to the well in the frequency tabulation.

The logical extension of Prelat's study is to use the probabilities obtained to forecast the outcomes of prospects not yet drilled. This could be

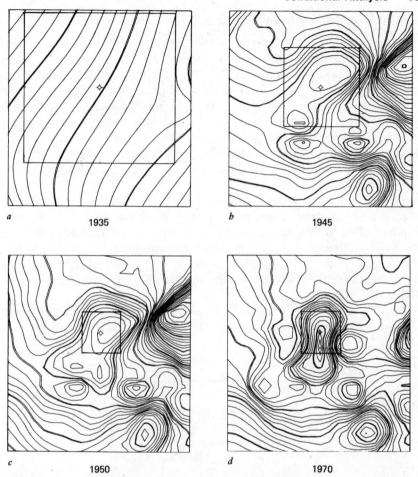

FIGURE 2.22. Reduction in cell size necessitated by progressive increase in complexity of interpretation as additional well information has become available. Overall area is 5 miles square. From Prelat (1974).

done in the area in which the probabilities were developed, or in an area where geologic controls of oil accumulation were judged to be similar. Although Prelat's study stopped short of this final objective, the methodology which he employed is directly applicable in other settings. The major shortcoming of his technique is the difficulty of consistently classifying structure by visual appraisal. Nevertheless, his method is a significant advance over most prospect evaluation methods because it attempts to objectively estimate outcome probabilities on the basis of the historical record of success.

Exxon's San Joaquin Valley Study

In the 1960s a team of Exxon researchers developed methods for handling large quantities of geologic and oil-production information. Their work (Grender, Rapoport, and Segers, 1974) involved the systematic collection of data in a test area near Bakersfield, California, in the southern part of the San Joaquin Valley (Figure 2.23). The study is particularly notable because it considered the distribution of geologic variables and oil production in three-dimensional space. The study area was divided into a succession of three-dimensional cells (Figure 2.24) for which a variety of numerical measures of geology and production were computed on a cell-by-cell basis.

Results of the study were expressed as relative frequencies, defined as the ratio of the number of times a feature occurred relative to the total number of observations. Relative frequencies can be used as probability estimates, although they were not treated from a probability standpoint in the Exxon study. Their work provides a framework of systematically assembled data from which a variety of probabilistic relationships can be extracted, however.

The Exxon test area extends over about 1000 square miles, and is part of a region that has been extensively explored. Oil fields in the area (Figure 2.25) are found in fault traps, noses and anticlines, hydrodynamic

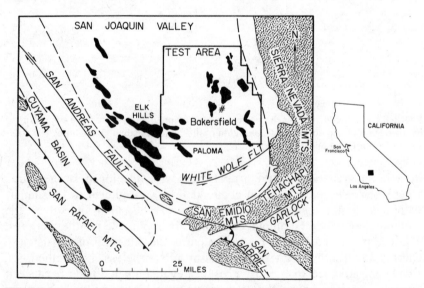

FIGURE 2.23. Exxon test area in San Joaquin Valley of California. Only larger oil fields are shown. From Grender, Rapoport, and Segers (1974).

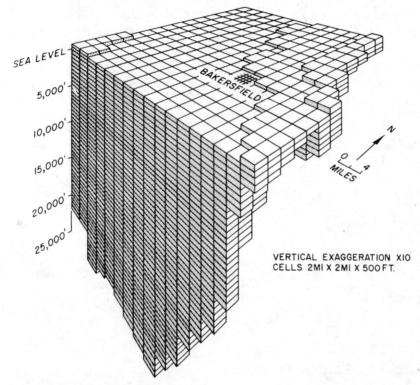

FIGURE 2.24. Block diagram showing partition of Exxon test area into subsurface cells. Each cell is 2 × 2 miles in geographic extent, and 500 feet thick. From Grender, Rapoport, and Segers (1974).

traps, and stratigraphic wedgeouts. Reservoir rocks include marine and nonmarine sandstones, fractured shale, and schist in the basement complex. One of the objectives of the study was to statistically relate oil occurrence to various geologic factors, including rock type, stratigraphic wedge position, and reservoir type.

The selection of cell size was a compromise between the desire to accommodate geological details, versus the need for economy in the number of cells to be handled. The area was divided into 249 vertical prisms each with an area of 4 square miles (2 miles × 2 miles). In turn the prisms were compartmented into 500-foot vertical increments, to yield a total of 5976 "unit cells" as shown in Figure 2.24. Each column extended from the ground surface to the top of the metamorphic basement complex. When necessary, unit cells were subdivided into thinner cells con-

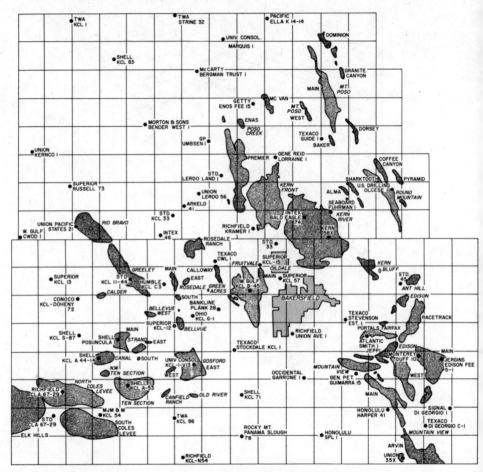

FIGURE 2.25. Oil fields and locations of key wells in Exxon test area. Grid squares are two miles on a side and each square specifies a column of cells as shown in Figure 2.24. From Grender, Rapoport, and Segers (1974).

sisting of vertical increments 5 feet thick, to ensure that thin but important reservoir units were appropriately represented in the data file.

The frequency data compiled in the study were related to several geologic parameters, ultimately represented as type and size of tectonic structure and as "stratigraphic wedge position." Tectonic structures were classified according to the scheme in Table 2.7 and are based on features in the generalized "form-line" map of the base of the Pliocene (Figure 2.26). Gentle folds consisting of homoclines, domes, noses, and synclines within the specified size categories were visually interpreted

TABLE 2.7. Two-way classification of structures based on form-line contour map in Exxon test area. Identifying codes (VSN = very small nose, etc.) are used in subsequent tables. From Grender, Rapoport, and Segers (1974).

	Size (in Square Miles)				
Type	Very Small 0—2	Small 2—4	Medium 4—16	Large 16—256	Very Large 256+
Homocline H	Not subdivided according to size				
Nose	VSN	SN	MN	LN	VLN
Dome	VSD	SD	MD	LD	VLD
Trough	VST	ST	MT	LT	VLT

from this map. A form-line structure map does not have a specific contour interval and represents the vertical relief of structures in only a qualitative or relative way. The reason that a form-line map was used rather than a conventional structure contour map is that the relief of the structures increases with depth. By using a form-line map, structures in the deeper western part of the area could be entered into the same tabulations as structures in the eastern part of the area. Other attributes of structures, including their location, orientation, approximate area, and gross configuration are correctly represented by a form-line map.

The other principal geologic parameter, stratigraphic wedge position, is a composite of both lithologic type and position with respect to large-scale sedimentary wedges. Four generalized lithofacies types were distinguished: coarse-grained marine, fine-grained marine, nonmarine, and basement. In turn, these lithofacies (except for basement) were classified by a system which takes their lateral and vertical succession (e.g., wedge position) into account. Wedge position (Figure 2.27) is classified either as a transgressive sequence, a basinward wedgeout, a regressive sequence, or a wedgeout of marine sands between nonmarine units.

In the present context, the most important aspect of the Exxon study is the ease by which relative frequencies of oil occurrence can be calculated. These frequencies can be determined so that they are jointly conditional on structure, lithology, and wedge position. Table 2.8 lists volumes in cubic miles according to structure type as defined in Table 2.7, and against lithology-wedge position as specified in Figure 2.27. Table 2.9 jointly lists recoverable oil reserves with respect to structure and lithology-wedge position.

Table 2.10 represents a composite of Tables 2.8 and 2.9 and expresses the recoverable reserves on a per-cubic-mile basis. We may call this the

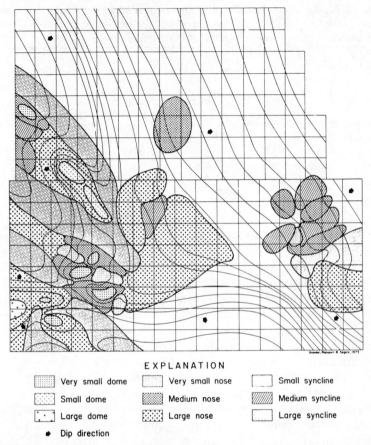

EXPLANATION

Very small dome	Very small nose	Small syncline
Small dome	Medium nose	Medium syncline
Large dome	Large nose	Large syncline
Dip direction		

FIGURE 2.26. Generalized "form-line" structure contour map showing geographic distribution of structural types specified in Table 2.7. From Grender, Rapoport, and Segers (1974).

"petroleum rating" and it may be expressed in millions of barrels of recoverable oil per cubic mile of sediment. As Table 2.10 indicates, certain combinations of structural type and lithology-wedge position have particularly high petroleum ratings. For example, the combination of a small dome and coarse-grained marine sediment in a basinward pinchout has a petroleum rating of 147.6 million barrels per cubic mile in the test area. Clearly, this combination is particularly favorable for oil occurrence. However, the volume of rock in the test area within this category is only about 1 cubic mile, and therefore its potential for further exploration seems limited.

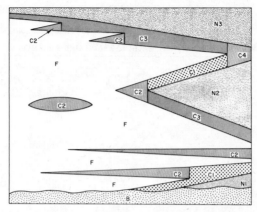

FIGURE 2.27. Hypothetical east-west stratigraphic section showing generalized stratigraphic relationships in San Joaquin Valley test area. Four generalized lithologic types are distinguished: coarse-grained marine (C), fine-grained marine (F), nonmarine (N), and basement (B). Vertical and lateral succession are represented by suffixes. Suffix "1" implies a transgressive sequence (e.g., in order of deposition, $N1$-$C1$-F), whereas regressive sequence is identified by suffix "3." Basinward pinchouts are indicated by a "2"; pinchouts of marine sands between nonmarine units are indicated by suffix "4." From Grender, Rapoport, and Segers (1974).

TABLE 2.8. Rock volume distribution in San Joaquin Valley test area classified jointly by structural type and lithology-wedge position. From Grender, Rapoport, and Segers (1974).

Structural Type	Lithology-Wedge Position							
	C1	C2	C3	C4	N1	N2	N3	F
H	62	90	103	19	20	8	329	368
VST								
ST	1	4	1					10
MT	4	7	4		1	1	5	18
LT	7	27	19		2		82	183
VLT								
VSN								
SN								
MN	4	12	2	1	1	1	5	35
LN	13	53	53	3		2	147	223
VLN								
VSD	1	6						21
SD	1	1						4
MD								
LD								
VLD								

TABLE 2.9. Distribution of recoverable reserves in millions of barrels in test area. From Grender, Rapoport, and Segers (1974).

Structural Type	Lithology-Wedge Position								
	C1	C2	C3	C4	N1	N2	N3	B	F
H	266.9	3.2	44.1	7.4		58.7	522.5		17.6
VST									
ST		16.8							
MT	5.6	8.8				3.4	1.5		2.3
LT		39.0					0.2	0.5	
VLT									
VSN									
SN									
MN	7.7	38.2	2.8	3.7		11.5	0.6	1.6	2.8
LN		326.5	67.4			88.0	75.2	51.0	
VLN									
VSD		94.8							
SD		147.6							
MD									
LD									
VLD									

TABLE 2.10. Petroleum ratings, in millions of barrels of recoverable oil per cubic mile of sediment, in test area. From Grender, Rapoport, and Segers (1974).

Structural Type	Lithology-Wedge Position							
	C1	C2	C3	C4	N1	N2	N3	F
H	4.3	Trace	0.4	0.4	0	7.3	1.7	0.1
VST								
ST	0	4.2	0					
MT	1.4	1.3	0		0	3.4	0.3	0.1
LT	0	1.4	0		0		Trace	0
VLT								
VSN								
SN								
MN	1.9	3.2	1.4	3.7	0	11.5	0.1	0.1
LN	0	6.2	1.3	0	0	44.0	0.5	0
VLN								
VSD	0	15.8			0			0
SD	0	147.6						0
MD								
LD								
VLD								

The San Joaquin Valley study by the Exxon group is notable in that it deals with data that permit the statistical interrelationships between interpretive geology and oil occurrence to be analyzed. Unfortunately, except for calculating petroleum ratings, little subsequent use is made of the information for statistical forecasting. For example, frequency distributions could be calculated and conditional probabilities estimated using a Bayesian approach. Furthermore, the data would permit calculation of the expected monetary value of oil to be discovered by exploration in which different combinations of geologic features are considered.

Use of Proximity Equations in Conditional Analysis of Well Logs

The data contained in geophysical well logs provide a rich source of information which can be used in oil-occurrence forecasting. Both geologists and log analysts have used information from well logs to make interpretations in which oil occurrence is conditional on geological and reservoir-rock characteristics.

Breitenbach and Peterson (1972) have applied statistical methods in an attempt to extend the analytical capabilities of the geologist and log analyst. Their technique involves calculation of empirical "proximity equations" which relate distance from an oil field to well-log data. As many potentially useful variables as possible are extracted from digitized log data. Typically, these variables include structural elevation and thickness of stratigraphic units, shale resistivities, shale gamma radiation levels, and many other properties which may bring the total number of variables to several hundred. These variables are analyzed by stepwise multiple regression to yield an expression relating them to distance to known oil production.

In the regression, distance to known production is considered as a dependent variable, or variable to be estimated, and the various well-log properties are the independent or predictor variables. The statistical relationship between the dependent variable and the most effective (in the sense of best predicting) of the independent variables is determined first. Then the next most effective variable is added to the equation, and so on. The process continues until the most effective set of variables has been found, and the remaining independent variables that do not contribute to the estimation of proximity from production are omitted. The stepwise procedure reduces the extremely large set of variables down to a smaller number of variables which are the best predictors of proximity to oil. Unfortunately, Breitenbach and Peterson do not provide information on the proportion of the total variance accounted for by the regression equation. The proportion left unaccounted for may be critical in evaluating their results.

Once the coefficients of the regression or proximity equation have been found, the estimated proximity to oil can be calculated for all locations in an area and the points can be contoured. The resulting map is a statistical forecast that is conditional on the combination of geological variables extracted from well logs. This forecast could be cast in probabilistic terms by considering the error variance of the predicting equation. However, this requires assumptions about the statistical nature of the variables (primarily that they are normally distributed) that may be difficult to justify. Breitenbach and Peterson apparently stop short of such probability estimations.

The Future of Conditional Analysis

Conditional analysis will assume increasing importance in oil exploration in the future. The remainder of this book is devoted to the current state of its application to the search for petroleum. Examples that follow are drawn largely from the work of our associates and from our own research; very little else has been published on conditional analysis for exploratory purposes. Such work, of course, is highly proprietary so that the relative dearth of published information may not reflect the degree of interest nor activity in this field.

We emphasize that the specific examples which follow do not demonstrate the last word in the methodology of conditional analysis. Rather, these are starting points from which more elaborate and effective procedures can be developed by those who have the necessary data files and financial incentives. An immediate challenge is to apply these approaches in geophysical exploration. We have not dealt with geophysical information here because it is not available in the areas where our test cases have been located, but there is no reason why geophysical variables cannot be effectively analyzed by conditional methods. The ultimate objective of conditional analysis is to combine *all* relevant information, from whatever source, into an integrated whole. Even then, exploration will remain a probabilistic endeavor, but one in which the odds are known with greater certainty and the risks can be seen more clearly.

If a reliable estimate of the frequency distribution of oil-field sizes is available for a region, the probabilities attached to the discovery of fields of different magnitudes can be easily calculated. Probabilities obtained in this manner have only limited use, however, because they are independent of any considerations of local geology and have no direct application in appraising individual prospects. Frequency distributions for a region are useful in evaluating the potential of large areas and can be used to establish general exploration policy. In this book, we do not deal further with such estimates, as our main interest is methods of appraising the outcomes of individual plays and prospects. The difference is partly one of scale; we are concerned with statistical and probabilistic applications at the detailed level which must concern the explorationist as well as the corporate manager.

Objective tabulation of frequencies requires that data be treated systematically; thus like features are classed together. Most geological properties that are used as guides to the occurrence of oil are conveniently expressed by contour maps, including subsurface structural maps, isopach maps, permeability maps, seismic reflection time maps, and so on. As we have seen, these maps can be analyzed to yield frequencies of occurrence of specific features and the outcomes of exploratory tests of these features. It is essential, however, that the maps used to obtain these frequencies be created in a consistent manner so differences in contouring methods do not confuse the results. The next chapter provides a comprehensive review of computer methods for constructing contour maps of geologic variables.

CHAPTER 3
Contouring by Computer

MACHINE VERSUS MANUAL CONTOURING

A computer program that will rapidly and accurately prepare contour maps of geologic data is essential in probabilistic exploration. Maps are not only the means by which conventional geologic plays are represented, but also serve as the basis for computations that eventually lead to probabilistic assessments of prospects. Although the relative merits of machine versus human mapping have been extensively argued since the development of the first computer contouring program, machines have slowly assumed a major role. This reflects the comparative ease and speed with which a computer can sort through massive data banks, a task which humans find especially onerous. However, speed and convenience are not the main reasons for the importance of automatic contouring procedures in the present context. Rather, computer contouring allows us to represent a geologic variable in a mathematical form that can be manipulated in ways that are not possible otherwise.

A contour map, whether drawn manually or by a plotter, depicts variation in a variable through space by the use of *isolines*, or lines of equal value. In the commonest application, the variable is elevation of a geologic horizon, and the isolines are contours that depict the form of a structural surface. The contour lines must "honor" points of known elevation, such as the well control used to make the map. That is, a contour line higher than a point must pass by that point on the "uphill" side of a slope, and conversely, a lower valued contour line must pass by the "downhill" side. Near data points this requirement constrains both man and machine. Between the known points, however, there is much more freedom for interpretation. A machine contouring program traces out contour lines by a precise mathematical relationship based on the geometry of the control points. A geologist, however, contours not only the control points but also his concepts and ideas about what the surface should look like. If these preconceived ideas are indeed correct, a competent geologist may be able to create a map which is superior to a machine-made product. On the other hand, if his preconceptions are erroneous, the finished map is likely to be seriously in error. An exploration geologist who is encouraged to "look for plays" may subconsciously succumb to the temptation to insert unsubstantiated closure between control points, rationalizing that the available data do not preclude the favorable possibility even if they do not obviously suggest its presence.

A more subtle form of bias may exist in man-made contour maps which are regularly updated during the course of an extensive drilling program. An initial map may be based on comparatively few control points, but still represents a significant investment in the time of geologists and draftsmen. As additional data become available, the file map is updated in a piecemeal fashion by erasing and adjusting contour lines to conform to the new information. Ultimately, the map becomes a patchwork of accommodations, but still reflects the ideas that went into the original map. Because geological concepts tend to evolve with time and additional information, this composite map may differ significantly from a new, redrawn map based on the same data.

The extent of the adjustment made for a new control point in a manually produced map depends on the whims of the draftsman and the closeness of other control points. An updated machine-made contour map, in contrast, is recreated in its entirety. The computer cannot recall its previous interpretations of the surface and so begins afresh. New data points are incorporated into the revised map in exactly the same manner as the original data. If these new points imply a radical revision of the form of the map, this is done without qualms.

In other words, a computer program is blindly consistent in the way it constructs a map. This is an essential virtue in the type of probabilistic analysis used here, because this requires repeated mapping of historical sets of data. Old maps dredged from the vault cannot be used, because earlier draftsmen or geologists may have interpreted an area in an entirely different manner than later workers. The same problem arises when attempting to compare two different mapped areas. How much of the difference is real, and how much is the result of the stylistic differences between different map makers?

However, the strongest argument for machine contouring is that it creates a mathematical model of the mapped surface that can be used for further analysis. In the process of plotting a map, a computer must first estimate the elevations of a tremendous number of intermediate points from the known control points. These intermediate points are usually on a regular grid and can be displayed directly as in Figure 3.1. This mesh of points provides the numerical values necessary for probabilistic interpretation, and cannot be obtained in any simple way from hand-drawn contour maps. In fact, the only manner in which they could be obtained requires digitizing the manually produced map, and then running the contour line values through the first phase of an automatic contouring program. It seems simpler to skip the manual step altogether.

Dahlberg (1975) has offered some reassuring insight into the nature of computer-drawn maps for those who regard them with suspicion. In an experiment, he pitted experienced petroleum geologists against a widely used automatic contouring program. The test data consisted of structural elevations from a collection of wells drilled into and around a Devonian reef in Alberta. Information from only a small number of the wells actually available was presented to the participants, and the objective was to assess the relative capabilities of men and machine in creating a realistic structural contour map. All the maps tended to be very much alike at and near control points, but differed radically in uncontrolled areas. Some geologists produced "better" structural maps than the computer program, in the sense that their representations were closer to the structural configuration revealed by the complete data set. Other geologists, however, were seriously in error. The interesting thing is that

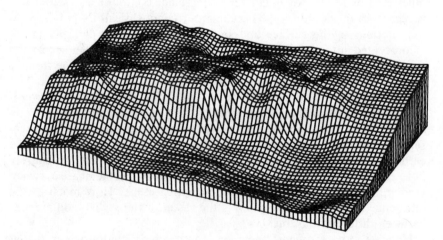

FIGURE 3.1. Perspective block or "fishnet" diagram of a subsurface structural horizon. Each intersection represents one grid point on structural surface whose elevation has been calculated by contouring program (Sampson, 1975b).

the automatically contoured map coincided almost exactly with the average of the manually produced maps. That is, between control points some geologists tended to bend their contour lines in one direction while others bent them in the opposite direction. Most drew their lines through a common middle ground, and only a few seriously deflected their contour lines one way or another. The computer drew its contours almost precisely through the middle of this bundle of lines. In this instance, the computer program behaved like an "average" geologist. As we shall see from the way contouring programs operate, this is not an unexpected result.

ASSUMPTIONS MADE IN AUTOMATIC CONTOURING

Several assumptions are embodied in a computer algorithm to create a contour map of a surface which has been measured at scattered points. The completed map reflects these assumptions and is a reasonable representation of the mapped surface only if they are valid. In general, a contouring program is designed to map a surface which is (a) single-valued at a point, (b) continuous everywhere within the map area, and (c) autocorrelated over a distance greater than the typical spacing between control points. These requirements are discussed below.

If there is only one possible value for a variable at a specific geographic location, that variable is single-valued. Examples include elevation measurements of ground topography and subsurface structure. There is no uncertainty associated with these measurements except for that arising from measurement error. Only in very unusual circumstances, such as in recumbent folds or low-angle thrust faults, can the surface assume more than one value at a location.

Some important geological variables are not so obviously single-valued. Measurements of porosity, for example, are statistical in nature, and repeated sampling and analysis at a single location may result in a suite of values. This results both from errors in measurement and from random, small-scale variations in the small samples of rock that are analyzed. Most automatic contouring programs cannot accommodate such repeated data, although it may be possible to reduce multiple observations to a single, representative value such as the average or mean which can then be mapped.

Automatic contouring procedures involve interpolation between control points. Because of the methods involved, all the points obtained by interpolation lie on a continuous, sloping surface between the control points. If the actual surface contains discontinuities such as faults, these will not be recognized by the contouring program but will be mapped simply as areas of steep slopes. Faults or discontinuities that are indicated in advance of mapping can be accommodated by arbitrary procedures which in effect insert boundaries into the map. The mapping program will draw the surface on opposite sides of boundaries as though they were entirely separate maps. However, it has not been possible to create a contouring algorithm which automatically recognizes unidentified faults or breaks in a surface.

Spatial Autocorrelation

A mapped variable is said to be autocorrelated if the value at a point is closely related to nearby points. Automatic contouring programs assume the variable being mapped is autocorrelated, because they operate by selecting all the nearest control points around a point to be evaluated, and then estimating that point as some type of average. If the surface is highly autocorrelated, all of these neighboring control points will have approximately the same value, and their average will be a reasonable estimate for an intermediate location (Figure 3.2). In contrast, if the surface is poorly

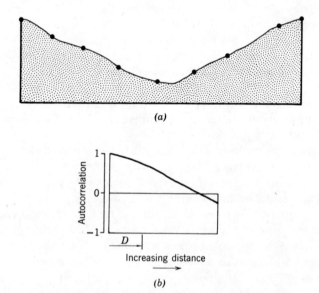

FIGURE 3.2. (*a*) Cross-section of highly autocorrelated surface. Sampling points where elevation is known are shown by large dots. (*b*) Autocorrelation function of surface. Length of arrow *D* indicates average distance between sampling points. All locations on surface are autocorrelated with known control points.

autocorrelated, neighboring control points will have little relation to one another, nor will they be related to the value at the point to be estimated (Figure 3.3). Under such conditions it may be impossible to make reasonable guesses about the nature of the surface between control points. In fact, the total absence of autocorrelation means a surface does not exist at all, because it implies that adjacent points may be completely dissimilar. Obviously, immediately adjacent points cannot be erratically different and still lie on a continuous surface.

Machine contouring procedures assume that closely spaced points are strongly related, but the degree of relationship lessens as the distance between points becomes greater. In more technical terms, the autocorrelation (which is a measure of the average similarity of pairs of points separated by specified distances) drops from high values for points that are closely spaced to nearly zero with increasing separation. When the distance between points becomes so great that there is no relationship between their values, it becomes impossible to combine them in any meaningful way to estimate the value of the surface at intervening locations. Thus, the spacing of control points is critical. If the data are so diffuse and scattered that there is no relationship between pairs of adja-

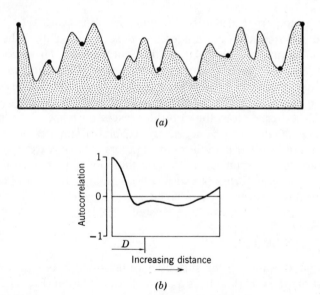

(a)

(b)

FIGURE 3.3. (a) Cross-section of surface with low autocorrelation; sampling points (dots) are at same locations as in Figure 3.2. (b) Autocorrelation function of surface. Length of arrow D indicates average distance between sampling points. Some locations are beyond influence of known control points.

cent points, there is no way of determining the form of the surface being mapped between the control points. This is true whether the map is made by machine or by hand.

Most automatic contouring programs also assume that the degree of autocorrelation of a surface is the same in all directions. That is, the surface being mapped is isotropic or without "grain." There are geological surfaces in which this is not true. A folded surface, for example, exhibits higher autocorrelation parallel to the fold than across the fold (Figure 3.4). A much better estimate of the elevation of a point on the crest of a fold can be made from control points which are also on the crest, though they may be quite distant. Closer control points lying on the limbs or in adjacent troughs are less reliable guides. It would be possible to design a contouring program that attached greatest importance to control points that lie along the "grain" of a data set. Unfortunately, such a program would have to be "custom-tailored" for every surface and would require a detailed analysis of the surface autocorrelation before mapping could be done.

HOW CONTOURING PROGRAMS WORK

In order to construct a map by computer, it is necessary to calculate a sequence of numerical coordinates that describe the path to be taken by the pen of a plotter which will draw the contour lines. These coordinates, which are accompanied by codes for raising and lowering the pen, are either transmitted directly from the computer to the plotting device, or are written on magnetic tape or punched in cards that are read by an off-line plotter. However, before the coordinate sets corresponding to contour lines can be calculated, it is necessary to obtain estimates of the mapped variable at points other than the control points. The only exception occurs in the specialized circumstances where the control points form a closely spaced regular grid across the surface to be mapped. General reviews of contouring procedures are contained in Walters (1969), Palmer (1969), Crain (1970), and Nordbeck and Rystedt (1972).

Triangulation Method

The estimated points on the surface may be found in one of three ways. The most obvious method uses a form of triangulation (Figure 3.5). Data points are connected by straight lines, forming a mesh of triangles that covers the map area. The triangles are constructed so that every control point is at the vertex of three triangles which are as nearly equilateral as possible. Points where specified contour lines cross the edges of these

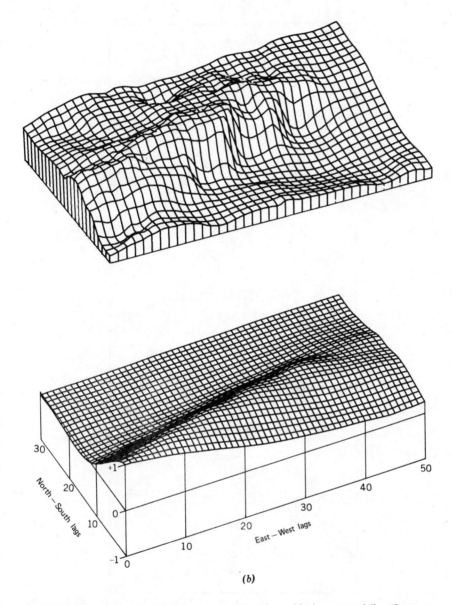

(b)

FIGURE 3.4. (*a*) Strongly anisotropic structural surface with elongate anticline (Sampson, 1975*b*). (*b*) Autocorrelation function of surface. Autocorrelation function measured parallel to anticline does not fall off rapidly with distance as does autocorrelation function measured perpendicular to fold.

97

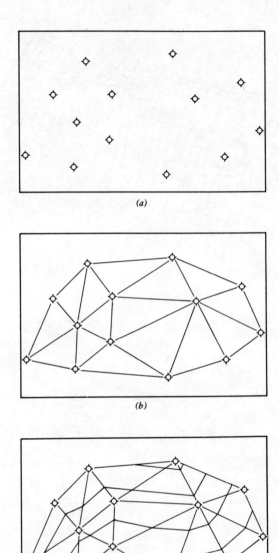

FIGURE 3.5. Triangulation method of estimating positions of contour lines. (*a*) Irregularly spaced data points. (*b*) Triangles formed across map area with data points as vertices. (*c*) Contour lines drawn through sides of triangles where points of specified elevation have been found by linear interpolation (after IBM, 1965).

triangles are determined by interpolating down the edges of the triangles between the control points at the vertices. This method is similar to triangulation by land surveyors and was one of the first contouring methods to be implemented on a computer (IBM, 1965). An elaboration on the procedure subdivides the triangles into sets of smaller, similar triangles by connecting corresponding points on opposite sides of the original triangles. The values at each vertex of the smaller triangles are estimated by fitting a low-order polynomial equation or trend surface by least-squares methods to the control points at the vertices of the major triangle and the adjacent three triangles. Contour lines are traced through this finer triangular mesh by interpolation down the edges of the small triangles. By successive subdivision of the basic triangular network, progressively smoother contour lines can be created.

The triangulation method has two drawbacks which have led to its almost total abandonment. First, the triangular mesh connecting the control points is not unique. A slight change in the way the points are connected will alter the path of contour lines across the map, sometimes in a radical fashion. Iterative procedures have been designed to create an optimal mesh, for example, by selecting most nearly equilateral triangles or by calculating Thiessen polygons (Rhynsburger, 1973) and using these to generate Delaunay triangles (Boots, 1974). However, these techniques may be sensitive to the arrangement of the data points and even to the order in which the data enter the computer.

A more severe restriction, however, is the computational inefficiency of the method. The geographic location of each control point, in the form of an X and Y coordinate, must be used explicitly through computations. The intersections of contour lines with sides of triangles must also be used explicitly. As a consequence, contouring by triangulation commonly uses many times the amount of computer time required by the more widely used gridding programs.

In addition, triangulation methods do not lend themselves to probabilistic map analysis of the type considered here. Later, we show that it is necessary to be able to directly compare contour maps constructed using different control points. Since the triangular mesh uses the control points as its nodes, obviously the meshes for two different maps will not coincide unless they are based on identically placed points. Thus, even disregarding the practical disadvantages of the triangulation method, it is inappropriate for our purposes.

Global Fit Method

A radically different approach to automatic contouring is the global fit method. The technique is an extension of curve-fitting procedures widely

used by engineers, and so it quickly attracted the attention of early
workers in computer graphics. These methods are called "global" in
contrast to "local" because all control points in the map area are considered simultaneously. An attempt is made to fit some type of involved
mathematical function to these points, usually by least-squares methods.
This approach has the computational advantage that sorting and searching
to find nearby sets of control points are not necessary. Once the global
function is found it can be used to estimate the value at any location in the
map area without further reference to the control points.

Global fitting is most familiar as trend surface analysis, where a
polynomial expansion of the geographic coordinates of the control points
is fitted to values of the mapped variable. In general form the equation for
a plane is

$$Z_i = \alpha + \beta_1 X_i + \beta_2 Y_i$$

where Z_i is the elevation of the surface being mapped at control point i, X_i
and Y_i are X and Y coordinates of points i, and α, β_1, and β_2 are
coefficients to be determined. These coefficients are found by the method
of least squares, which ensures that the quantity

$$\sum (\hat{Z}_i - Z_i)^2$$

is a minimum. $\hat{Z}_i$ is the estimated value of Z_i, calculated from the fitted
equation. This indicates the principal drawback of global methods as a
way of producing contour maps; the surface defined by the fitted equation
does not go through the control points. Rather it is a smoothed and
subdued approximation (Figure 3.6).

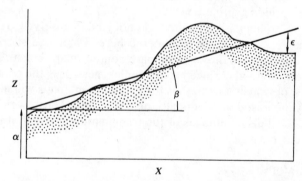

FIGURE 3.6. Approximation of surface (represented in cross-section) by simple equation
of form $Z = \alpha + \beta_1 X + \beta_2 Y + \epsilon$. Difference between true elevation and estimate along global
fitted line is ϵ.

The complexity of the global function may be increased by adding additional terms, commonly higher polynomial powers or trigonometric functions of the geographic coordinates. Each additional term gives the fitted surface one additional degree of freedom, allowing another reversal of slope (Figure 3.7). Most geologic surfaces, however, are too complex to be represented by any single equation that is feasible to use. As we shall see later, trend surfaces have their uses in probabilistic exploration, but the production of fine resolution contour maps is not one of them.

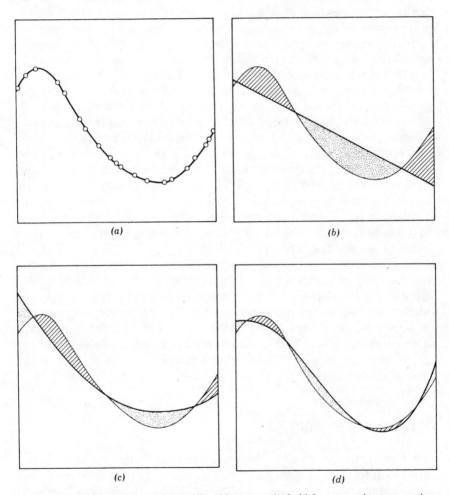

FIGURE 3.7. Global fits of polynomials with progressively higher terms, in cross-section. (*a*) Collection of original data points and line on which they lie. (*b*) Straight-line fitted to original data. (*c*) Parabola fitted to data. (*d*) Cubic curve fitted to data. Shadings represent positive and negative residuals. (Davis, 1973.)

Extensive discussions of trend surface analysis are included in many texts on mathematical geology, including Krumbein and Graybill (1965), Davis (1973), and Agterberg (1974).

Local Fit Method

The most widely used contouring technique is the local fit method, which considers only control points selected locally from within a small neighborhood around a point to be estimated. The procedure is repeatedly applied across the map area until the entire map is covered by a regular grid or mesh of estimated points. Once the regular grid of estimates has been constructed, contour lines can be laced rapidly through this numerical array (Figure 3.8). Interpolating from the scattered collection of irregularly spaced control points to a regular grid of estimated values accomplishes several significant computational objectives. First, the step greatly reduces the data storage requirements of the program because the interpolated grid can be stored as an array of numbers. It is not necessary to retain the explicit X and Y locations of the grid nodes because they are implied by their row and column position in the array. Second, a vast reduction in computation time can be achieved, even though many intermediate points must be calculated. This is because all sorting and searching operations necessary to trace contour lines through the array can be done using very simple logic operations. These practical advantages are so overwhelming that almost all commercial contouring systems use an intermediate gridding step.

Once the array of estimated points has been determined, the accuracy and reliability of a map is fixed. Contour lines may be altered or embellished as they are drawn through the regular array but such changes are purely cosmetic. Smoothing the contour lines, for example, may improve the appearance of the finished map but cannot affect its accuracy except in an adverse manner. Map accuracy is a result solely of the manner in which the array or grid of interpolated points is constructed from the original control points.

OPTIONS IN GRIDDING

Gridding, or the calculation of the regular array of intermediate points, involves three essential steps. First, we must sort the control points according to their geographic coordinates. Second, from the sorted files, the control points surrounding a grid node to be estimated must be searched out. Third, the program must estimate the value of that grid node by some mathematical function of these neighboring control points.

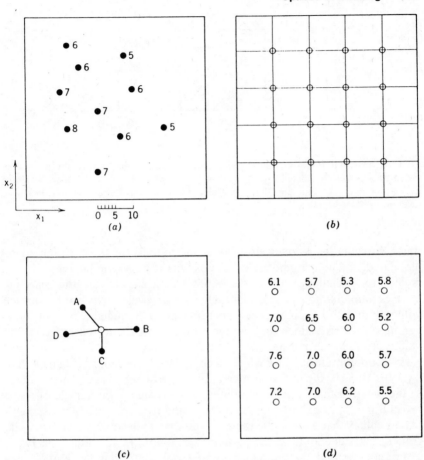

FIGURE 3.8. Steps in computation of grid nodes in contouring. (*a*) Original set of irregularly spaced control points on map. Numbers represent elevations. (*b*) Regularly spaced grid network with nodes to be calculated. (*c*) Location of four nearest control points with respect to particular grid node. Values at these four control points will be used to calculate estimated value at grid node. (*d*) Complete grid with elevation estimated at every node. (Davis, 1973.)

Sorting greatly affects the speed of operation, and hence the cost of using a contouring program. However, it has no effect on the accuracy of the estimates, so we will not consider it further. Both the search procedure and the mathematical function do have significant effects on the form of the final map.

The most obvious function that could be used to estimate a point on a map from nearby control points is simply to calculate an average of these points as shown in Figure 3.9 (Davis, 1973). In effect, this projects all of

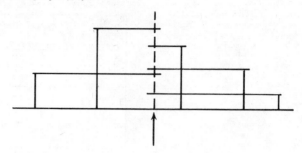

FIGURE 3.9. Interpolation from irregularly spaced data points to grid by method of weighted averages. Estimated value at grid node (arrow) is obtained as distance-weighted average of nearby points. In effect, control point values are projected horizontally to location to be estimated.

these surrounding known elevations horizontally to the location to be estimated. Then, a composite estimate is made by averaging these. If this is done on a regular grid over the entire map area, the resulting map will have certain characteristics. The highest and lowest areas on the surface will contain control points, and most interpolated values will lie at intervening elevations, since an average cannot be outside the range of the values from which it was calculated.

An estimation procedure which is considered to be more elegant calculates the slope or dip of the surface at the control points and projects these to the location that is to be estimated. An average is then made of these projections (Figure 3.10). Using this method it is possible to create estimated values that exceed the range of the data themselves, which may be good or bad depending on the particular situation. Incidentally, this is the estimation method used in many commercial contouring systems; it was originally devised by Osborn (1967) and has been described by Jones (1971) and Walters (1969). A similar two-stage procedure was developed by Batcha and Reese (1964) and is also used in several commercial systems (IBM, 1965).

Weighting Functions

The control points used in estimating a grid node, whether they are projected or not, are ordinarily weighted. The weightings assigned usually vary according to the distances between the location being estimated and the control points. Figure 3.11*a* is a location map of irregularly spaced wells around a point that represents part of a regular grid. The elevation of a formation top has been measured in each of these wells and is to be used to prepare a structure contour map. A weighting function that declines as

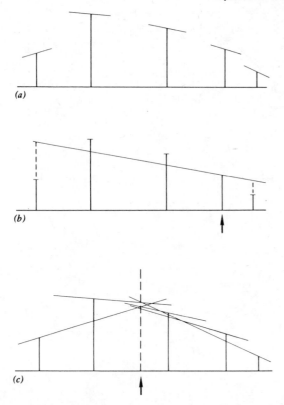

FIGURE 3.10. (*a*) Interpolation from irregularly spaced data points to grid by dip-projection method requires calculation of local dip at every data point. (*b*) Local dip of surface at data point (shown by arrow) is found by fitting least-squares plane to surrounding points. Fitted plane is constrained to pass through the data point. (*c*) Local dips at control points are projected to grid node (arrow). Value assigned to node is weighted average of these projections (Sampson, 1975*a*).

the inverse of the distance between the control wells and the location being estimated is shown in Figure 3.11*b*. If this weighting function is superimposed on the map in the form of the concentric contour lines representing the weights assigned to the control points, we see that comparatively distant wells may have a significant influence on the structural elevation that will be assigned to the grid node.

Most automatic contouring programs use a weighting function that drops off more rapidly than the inverse of the distance. A commonly used function is the inverse of the square of the distance, which might be called a "gravity" function. Figure 3.12 contains a graph of $1/D^2$ superimposed

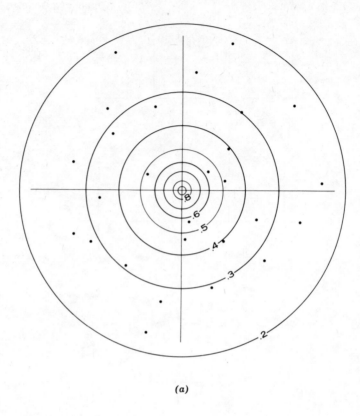

(a)

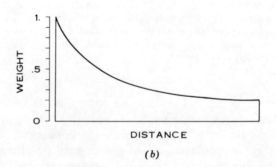

(b)

FIGURE 3.11. Influence of control points on estimate of grid node using inverse-distance weighting function. (a) Map showing location of irregularly distributed well locations used to estimate structural surface at grid intersection. Concentric circles are contour lines that denote relative weights attached to wells according to distance from location being estimated. (b) Graph of weight assigned control wells as function of distance from estimated location.

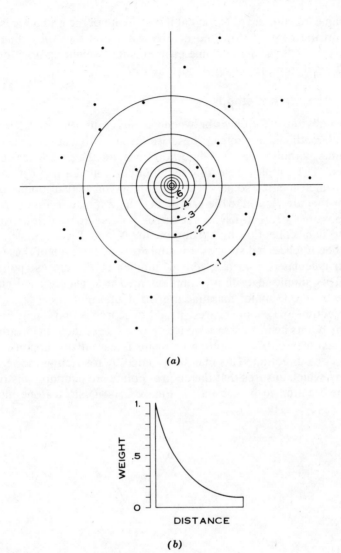

(a)

(b)

FIGURE 3.12. Control point influence with inverse distance-squared weighting. (*a*) Map showing well locations with concentric contours of weighting function superimposed. (*b*) Graph of inverse distance-squared weighting function.

on the same location map. Note that the estimate of the grid node is based mostly on the heavily weighted nearby wells and that the influence of more distant wells is slight. Some programs use weighting functions that drop off even more rapidly with distance.

Data-Point Search Methods

The most obvious differences between various contouring packages are in the search methods employed. These are the algorithms used to select the data points within a local neighborhood around the grid location to be estimated. The simplest selection technique is called a nearest-neighbor search, which locates some specified number of control points or well locations that are closest to the grid node being estimated (Figure 3.13). A set of possible nearest-neighbor control points are selected from the complete data collection by sorting on the X and Y coordinates of the points. The Euclidean distances from the grid node to each of these points are then calculated, and a specified number of the closest points are found. This simple algorithm is implemented in a widely used package distributed by one major manufacturer of digital plotters.

An objection to a simple nearest-neighbor search is that it may find that all nearby points lie in a narrow wedge on one side of the grid location that is to be estimated. The resulting estimate is essentially unconstrained, except in one direction. This may be avoided by restricting the search in some way which ensures that the control points are equitably distributed about the location to be estimated. Figure 3.14 illustrates one mode of

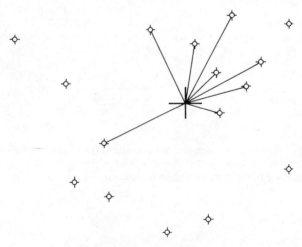

FIGURE 3.13.   Search technique which locates n nearest neighbors around grid node being estimated. No constraints are placed on radial distribution of control points.

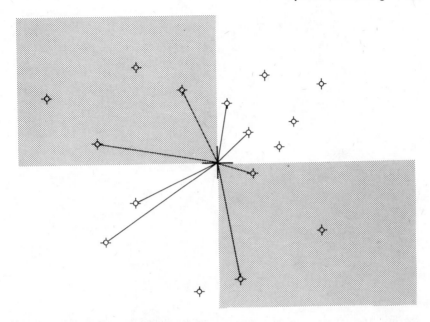

FIGURE 3.14. Quadrant search pattern used to insure equitable distribution of control points around grid node being estimated.

radial constraint, called a quadrant search. Some minimum number of control points must be taken from each of the four quadrants around the grid node being calculated. An elaboration on the quadrant search is an octant search, shown in Figure 3.15, which introduces a further constraint on the radial distribution of the points used in the estimating equation. A specified number of control points must be found in each of the octants surrounding the grid node being estimated. This search method is one of the more elegant procedures currently employed and is used in several proprietary programs, including a popular one sold by a major computer manufacturer (Walters, 1969).

Any constraints on a nearest-neighbor search, such as a quadrant or octant requirement, will obviously expand the size of the neighborhood around the location being estimated. This is because some nearby control points are likely to be passed over in favor of more distant points in order to satisfy restrictions on the number of points that may be taken from a sector. Unfortunately, the autocorrelation of a typical geological surface decreases with increasing distance, so these more remote control points are less closely related to the location being estimated. This means the estimate may be poorer than if a simple nearest-neighbor search procedure were used.

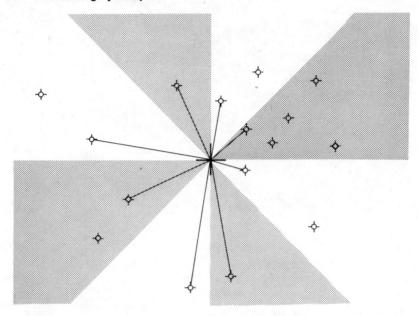

FIGURE 3.15. Octant constraint on search pattern around grid node being estimated.

Certainly, radial search constraints may be useful with unusual point distributions, as, for example, widely spaced traverses of closely spaced points (Figure 3.16). Such data point distributions are particularly apt to arise with seismic or other forms of geophysical information. It is common knowledge that machine contouring procedures find such data difficult to map in a convincing manner, as the programs often tend to create a pattern of alternating hills and hollows resembling a contour map of a gigantic egg carton. However, for more irregularly distributed control such as well data, simple nearest-neighbor search techniques do at least as well as more elaborate procedures.

AN EMPIRICAL TEST OF GRIDDING METHODS

An empirical analysis was performed to evaluate the differences between some of the more widely used variants of the local fit procedure (Davis,

FIGURE 3.16. Estimation of grid node from data on seismic traverse. (*a*) Nearest-neighbor search may select all points from single line, resulting in estimate which is unconstrained perpendicular to line. (*b*) Octant search constraint forces algorithm to select some points from adjacent traverses.

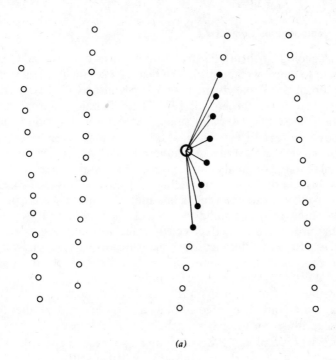

(a)

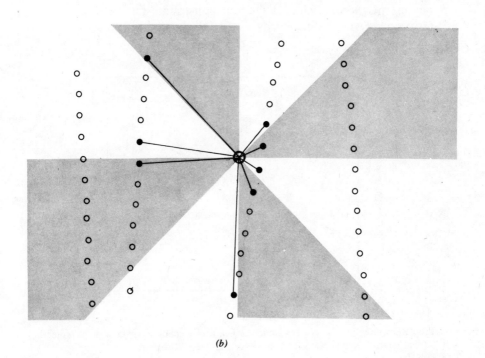

(b)

1976). A geologic subsurface structure map was constructed in an area of moderate to dense well control. The test data set includes all wells drilled to the top of the Pennsylvanian Lansing-Kansas City Group in Graham County, Kansas to the end of 1974. This includes approximately 3000 wells in an area 30 × 30 miles square. Structural elevation values are in feet above sea level, so statistics of errors are also given in feet. The data set is described more fully in Chapter 5. The fidelity of the computer-drawn map to the original control points was found by back-calculating to each well location from surrounding grid nodes. The root mean square (RMS) error of these differences is a measure of the scatter in values of the contoured surface around the true surface. A large RMS value indicates the procedure is ineffective, or inaccurate at the data points. The skewness of these differences is a measure of bias, or tendency for the surface representation to consistently fall above or below the correct values. Wren (1975) performed a similar analysis of contouring procedures using a set of aeromagnetic data from western Canada. However, he regarded a contouring algorithm as a two-dimensional filter and compared algorithms on the basis of their filter responses.

Figure 3.17 is a plot of RMS error and skewness of control-point errors for maps constructed by an algorithm which used various numbers of nearest neighbors, selected without constraints on the search pattern. Grid nodes were estimated by averaging the control points found, after

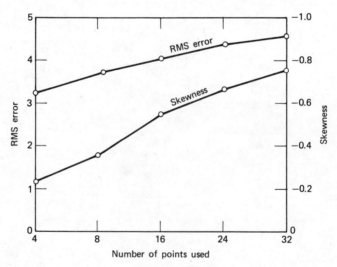

FIGURE 3.17.   Error at control points for nearest-neighbor searches without projection of dips at control points. RMS error is in feet. (Davis, 1976.)

weighting by a function which declines at the rate of $1/D^4$. Both RMS error and skewness increase as the number of nearest neighbors increases. Other plots for weighted projection methods and octant or quadrant search patterns are essentially identical to this illustration.

The influence of different weighting functions on control-point error is shown in Figure 3.18. Errors were found by back-calculating from a surface created using eight nearest neighbors, no search constraint, and no projection of dips. A function which is heavily influenced by nearby control points creates a surface representation that has the smallest RMS error and skewness. In contrast, a surface created using a slowly declining weighting function is smoothly undulating and has many of the averaging properties of a global-fit surface. Search constraints and projection of

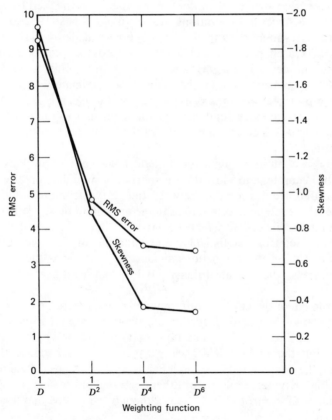

FIGURE 3.18. Errors at control points for nearest-neighbor searches for four different weighting functions, using eight points and without projection of surface dips at control points. RMS error is in feet. (Davis, 1976.)

dips do not significantly alter the degree to which the map representation honors the original control points.

The primary objective in many contour mapping exercises is, however, not to represent the available data as accurately as possible, but to estimate with minimum error values of the surface at locations where no control is available. In petroleum exploration, for example, structural contour maps are used to predict the locations of potential targets such as closed positive structures or anticlines prior to drilling. The ability of various algorithms to produce accurate estimates at locations where no control exists was checked by an empirical test using the same set of subsurface data. The well data were first divided into two subsets, one containing approximately 700 wells drilled prior to 1952, the other containing about 2700 wells drilled after that year. The set of early wells was used to generate structural contour maps that were checked by comparison with the structural elevations at the "blind," post-1952 locations.

Figure 3.19 summarizes the estimation errors made by various combinations of search patterns and numbers of control points used in the estimation process. The algorithm weights control points according to a function which drops off at the rate of approximately $1/D^4$, with or without dip projection. Bias is expressed by the mean error, which is a measure of the average tendency for the algorithm to underestimate or overestimate. RMS error is a measure of the inefficiency of the estimating procedure.

Ideally, a contouring algorithm should have both a low bias (i.e., it should be accurate) and a low RMS error (i.e., it should be precise). There is little difference in precision between the various combinations considered, but there are large differences in the amount of bias. Methods using large numbers of nearest neighbors have less bias, because the estimate has the character of a statistical average and the law of large numbers is operating. Use of dip projections significantly increases the bias, especially in areas where control density is low and local gradients may be high.

An example of the effect of different weighting functions on estimation error is shown in Figure 3.20, for an algorithm which uses an octant search constraint. Weighting functions that drop off slowly have the lowest bias but the highest RMS error. Again, this is because they are assuming the characteristics of a global averaging process. Relationships in this plot are typical of those for other combinations of search constraint and number of control points used, although the scales may be shifted somewhat.

The distressing (although not surprising) conclusion from this empirical study is that the different objectives of the contouring procedures consid-

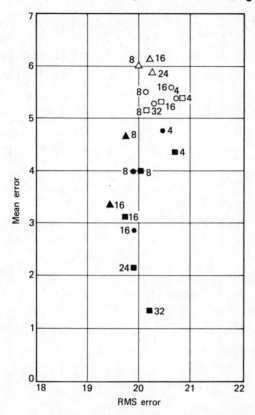

FIGURE 3.19. Errors at estimated points for different search patterns and numbers of points used in the gridding algorithm. Estimations are calculated as averages of control points weighted inversely to square of distance. Square symbols represent nearest-neighbor search, circles represent quadrant search, and triangles represent octant search patterns. Solid symbols represent methods without projection, open symbols represent methods with projection of dips. (Davis, 1976.)

ered are not mutually obtainable. An algorithm which faithfully honors the original control points should utilize a weighting function that drops off extremely rapidly with distance and which uses only a few nearest neighbors. However, such an algorithm will produce poor predictions or estimates at locations where no control is available. The best estimating procedure might be one that uses 16 or 24 control points in each calculation of a grid node, and which weights distant points relatively heavily. This, of course, would provide a poor reproduction of the original control points.

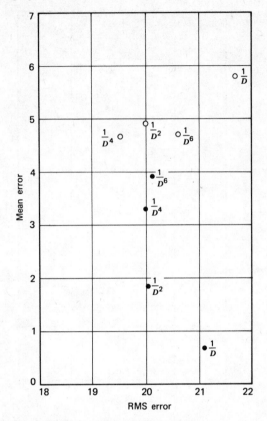

FIGURE 3.20. Errors at estimated points for different weighting functions, using octant search with two data points per octant. Solid symbols represent procedures without projection of surface dips, open symbols represent those with projection. (Davis, 1976.)

Since apparently it is not possible to specify a combination of features that will lead to a map that is "best" in both a representational and predictive sense, the selection of features should be based on the specific purpose for which an individual map is made. This requires a contouring package that contains a variety of alternative procedures, under the control of the user. There is, in addition, a third selection criterion, based on a practical constraint: Figure 3.21 shows variations in computation times with the number of nearest neighbors used by the algorithm. Since computing costs are closely related to processing time required, for routine or production contour mapping, run time may be as important as fidelity or predictive ability.

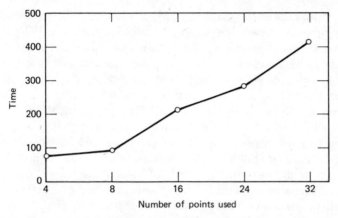

FIGURE 3.21. Run times for gridding phase of SURFACE II, in hundredths of seconds of Honeywell 635 CPU time, for nearest-neighbor searches without projection of dips at control points. Dip projection increases run times by constant factor equal to (number of data points)/(number of data points + number of grid nodes). (Davis, 1976.)

SURFACE II Contouring System

Most maps shown in this book were made with the SURFACE II contouring package developed by the Kansas Geological Survey (Sampson, 1975*b*). This is an extremely flexible program containing a wide variety of search pattern and weighting function options, which can be combined to emulate many of the currently used contouring programs. The package also contains extensive error-checking and analysis routines that allow comparative studies to be made of the effectiveness of different approaches (Davis and Sampson, 1974). In order to make the maps in this book comparable to those produced by commercial programs, a combination of options has been selected that are typical of programs used in the petroleum industry. All maps have been constructed using eight nearest neighbors, found by a search procedure that places no restrictions on their radial distribution. Surface dips at control points have been calculated and projected to the grid nodes to be estimated, where they have been averaged after weighting by a scheme which drops off somewhat more rapidly than $1/D^4$. It produces contour maps that are a reasonable compromise between the need for reliable prediction and the desire for accurate representation at the control points.

Automated contouring procedures are in a state of rapid evolution. The methods described here will probably be supplanted by modifications that are computationally more efficient and at the same time more accurate. For example, the method of surface estimation known as kriging creates optimal estimates of the mapped configuration by utilizing knowledge of the autocorrelation between points (for a review see Olea, 1974). Methods of automatically determining the nature of surface autocorrelation are presently underdeveloped. These methods can be combined in an automatic contouring package that would first determine and then use the best weighting functions for a particular data set. In effect, the contouring algorithm would be "custom-tailored" for each application.

Other advances may come in the development of nongridding contouring programs, perhaps using bicubic spline functions (Harder and Desmarais, 1972). The ability to use ancillary information such as dip meter readings may increase the accuracy of subsurface maps. Whatever the nature of these improvements, they will enhance our ability to make probabilistic estimates because they will reduce one of the fundamental sources of uncertainty in exploration.

CHAPTER 4

Local Configuration of Geologic Surfaces

MEASURING LOCAL CONFIGURATION OF A SURFACE

Explorationists are not so concerned with the absolute magnitude of a geologic variable as they are with its value at a point in relation to surrounding points. Thus, structure is characterized in relative terms such as anticline or syncline rather than in absolute distance above or below sea level. Similarly, permeability measurements may have limited significance in themselves, but local variations in permeability can be extremely important in defining exploration targets. It is not sufficient simply to map a geological variable across an area. We must also be able to define and measure the local shape of the mapped surface.

Unfortunately, it is difficult to quantify expressions of local shape such as an anticline. Attempts have been made to qualitatively define local structure within small areas or cells (Abry, 1973, 1975; Prelat, 1974; Grender, Rapoport, and Segers, 1974), but such definitions are subjective and hence liable to bias and inconsistency. Many ambiguous or borderline cases arise when attempting to classify the structure within a single cell without reference to surrounding cells, and it is difficult for a human to resolve these in a totally consistent manner.

Volume of Closure

One widely used direct measure of local structure is estimation of volume of closure, a technique employed in Chapter 6. At first glance, it would seem simple to calculate volume of closure for a machine-contoured map.

The operation is not performed on the map itself, but on the numerical array of interpolated values from which the map is drawn. First the array is searched to locate isolated highest points, which become the centers around which closure volumes will be calculated. Next, a search is made outward from these centers to determine the point at which the downward slope of the surface reverses. This defines the spill-over point. The volume of closure can be found by numerically integrating under the surface for all points higher than the spill-over elevation.

Unfortunately, this technique is difficult to apply in practice. In a map containing relatively large, simple structures, the spill-over points can be calculated with minimal ambiguity. More typically structures are found in a variety of sizes, with smaller structures superimposed on larger. It then becomes difficult to establish rules to determine which saddle points are to be chosen as the spill-over points, as in Figure 4.1. Computational problems become formidable as well, because the operation necessitates a hierarchy of iterative searches and comparisons. For a map array of even moderate dimensions this becomes extremely expensive.

Rate of Curvature

Another direct method of measuring local structure is to calculate the rates of curvature of the mapped surface. An extensive set of statistics for this purpose has been devised by Demirmen (1973a), and a procedure for their calculation directly from a contour map or numerical array has been programmed (Demirmen, 1972). Some of these measures are used to describe local configuration of a map of lithologic variation in a case history described later in this book.

Basically, the procedure operates by dividing the map area into a series of nonoverlapping cells of uniform size and shape. Profiles of the surface are defined along radial lines, called "polar lines," emanating from the centers of these areas (Figure 4.2). Demirmen defines four measures of the shape of these surface profiles. *Polar slope index 1* is the mean of slopes measured between successive points along a profile. These points are taken whenever the radial measurement line intersects a grid line in the numerical array of estimated values which define the mapped surface. This statistic, symbolically indicated by $\sigma_j^{(1)}$, is positive if, on the average, the surface along the profile dips toward the center of the cell and negative if the profile dips away. *Polar slope index 2,* designated $\sigma_j^{(2)}$, is equivalent except that the slope is calculated between the center of the cell and successive points along the profile. As Demirmen notes, there is an analytical relation between these two slope measures.

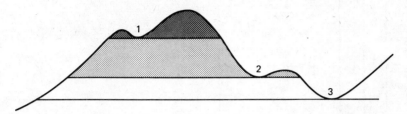

FIGURE 4.1. Cross-section of structural surface. Successive spill-over points occur at levels 1, 2, and 3.

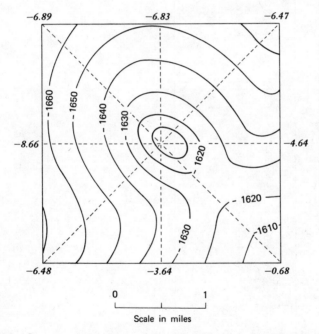

Scale in miles

FIGURE 4.2. Structure contour map of Pennsylvanian Viola Limestone in Wilmington Field, northeast Kansas, with radial pattern of superimposed "polar lines." Values in italics are marginal polar slopes along dashed lines. After Demirmen (1972).

TABLE 4.1. Shape parameters for the Wilmington field structure (Figure 4.2) tabulated for polar lines 1 through 8 (after Demirmen, 1972).

Shape Measures	Polar Lines							
	1	2	3	4	5	6	7	8
Polar slope index 1, $\sigma_j^{(1)}(\times 10^3)$	−3.086	−6.470	−6.812	−6.768	−7.963	−6.317	−3.251	0.887
Polar slope index 2, $\sigma_j^{(2)}(\times 10^3)$	−4.630	−7.410	−7.710	−7.690	−9.828	−9.300	−6.913	−2.259
Marginal polar slope, $\sigma_j(\times 10^3)$	−3.030	−6.472	−6.830	−6.885	−8.162	−6.478	−3.636	0.678
Polar curvature index, $\kappa_j(\times 10^6)$	4.709	0.981	0.897	0.785	3.160	8.304	11.957	9.076

The *marginal polar slope*, σ_j, is the slope of a straight line connecting the center of a cell to the point of intersection between a polar line and the cell margin. The *polar curvature index*, κ_j, is the ratio of the arc length to chord length between the cell center and the map margin along a polar line. This statistic is computed so that it is positive if the profile is concave-upward and negative if the profile is concave-downward.

Demirmen measured four statistics for each of eight polar lines, giving 32 measures of surface form in each cell (Figure 4.2 and Table 4.1). In a test of the procedure Demirmen (1973*b*) calculated shape parameters for subsurface structure in Stafford County, Kansas. The area was divided into 1265 small cells, each 1.5×1.5 miles square and each centered on an exploration well. The purpose of the study was to determine if there was a probabilistic relationship between the shape statistics generated according to this procedure and the occurrence of oil. An attempt was made to combine the polar slope statistics into composite measures of shape by principal components analysis. The attempt was not fruitful, however, and better discrimination between producing and dry wells was achieved using simple measures, such as the magnitude of trend surface residuals.

Aside from the obvious computational effort involved, this direct method has a serious drawback in many practical applications. It requires precise knowledge of the form of the surface being measured, but this information may not be available. Within a small cell, for example, there may be only a few control points where the elevation of the surface is known. All other elevations at grid nodes within the cell are estimated from these and adjacent control points. It is possible that the number of shape statistics calculated within a cell approaches or even exceeds the number of observations that define the shape of the surface within the area. In such a circumstance, the shape statistics must be highly redundant and will be only marginally reliable.

Filtering

Filtering is an indirect method of measuring the magnitude of local structures. The technique is an extension of methods used in geophysics and in electrical engineering to extract signals from a welter of background noise. Robinson (Robinson and Ellis, 1971; Robinson, 1975; Robinson, Charlesworth, and Ellis, 1969) has been a leading proponent of filtering techniques for the study of subsurface structural configuration, using filter analysis to analyze regional structural patterns in Alberta.

Although filtering is an application of double Fourier analysis, it can be considered simply as a form of template matching. The "template" or filter consists of a small matrix or array of numbers, arranged so that they coincide with the pattern of numbers in a map grid. The numbers within the small array are weights and may be positive or negative, but must sum to zero. There should be an odd number of rows and an odd number of columns in the filter, so that a central point exists.

The filter is cross-multiplied with the map grid to create a new array of filtered values. This new array is the same size as the original map, except that a few rows and columns along the edges of the map are lost because they cannot be filtered. The values in this new array reflect the degree of correspondence between the template and the surface being filtered.

In operation, the filter array is centered over successive points in the map grid. At each point, corresponding elements in the filter and in the map grid are multiplied. All of the resulting products are then summed and the total becomes the value in the filtered array, which corresponds to the location of the center of the filter. The operation can be visualized with the help of Figure 4.3, which shows a small 3 × 3 filter over part of a map grid. Part of the resulting filtered grid is shown below.

Figure 4.4 is a computer-contoured structure map of the base of the Fish Scales Sandstone in Alberta. This unit, with others, was filtered by

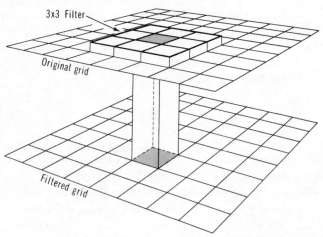

FIGURE 4.3. Diagrammatic representation of filtering process, using small 3 × 3 filter on part of map grid.

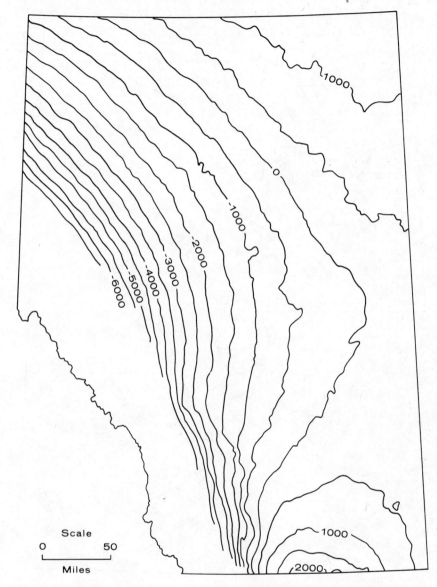

FIGURE 4.4. Structure contour map of base of Fish Scales Sandstone (Cretaceous) in Alberta. Contour interval is 500 feet. After Robinson, Charlesworth, and Ellis (1969).

FIGURE 4.5. Surface of basal Fish Scales Sandstone filtered to enhance features approximately 15 miles across. Ruled areas are positive structures having an amplitude greater than 100 feet. Unshaded closed areas have negative amplitudes greater than 100 feet. After Robinson, Charlesworth, and Ellis (1969).

Robinson, Charlesworth, and Ellis (1969) to isolate local structures. The base of the Fish Scales is a conspicuous Lower Cretaceous marker horizon, widely used for structural mapping in this area of Canada.

The structure contour map was originally prepared from almost 7000 wells. The raw well data were used to create a regular grid of estimated elevations on a 2-mile spacing (a preliminary study of the spatial frequencies present in the surface had been done at a quarter-mile spacing). A series of symmetric filters were used to isolate structures having sizes in the ranges 14 to 18 miles, 10 to 60 miles, and 18 to 100 miles. The output from the first filter is shown in Figure 4.5.

The pattern of local structures is interpreted by Robinson, Charlesworth, and Ellis (1969) as reflecting subdued folding that resulted from tectonic movements in the Cretaceous. The maps emphasize the existence of two sets of intermediate-sized structural trends, one NE—SW, the other NW—SE. The authors suggest that these penecontemporaneous structures may have been important in the control of many sedimentological patterns and may have influenced the location of important oil and gas reservoirs.

In this example, a nondirectional filter was used. That is, it does not

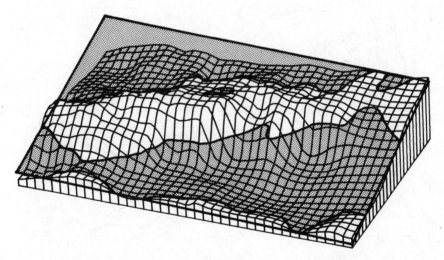

FIGURE 4.6. Block diagram of subsurface structural horizon and fitted first-degree trend surface. Shaded areas are below trend surface and form negative residuals; unshaded areas are above trend surface and form positive residuals.

preferentially accentuate elongated features. Such capability can be built into a template if desired and will enhance structures that possess one dimension greater than another. However, such filters are sensitive to the orientation of the map features, and they may actually subdue structures that are not aligned parallel to the major axis of the filter.

Trend Surface Analysis

Trend surface residuals also constitute a form of indirect measure of local structure, and are widely used in petroleum exploration for this purpose (Harbaugh and Merriam, 1968). Trend surfaces generally involve low-degree polynomial functions of the geographic coordinates of control

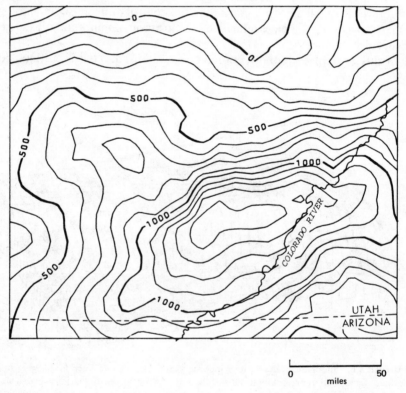

FIGURE 4.7. Isopach map of Cedar Mesa Sandstone (Permian) in Four Corners area. Contours in feet. Adapted from Irwin (1971).

points, fitted by least-squares methods to values at the points. The technique is a three-dimensional extension of curve-fitting procedures used in engineering and of regression methods used in statistics.

The trend surface equation describing a plane is a linear function of the form

$$Z_i = \alpha + \beta_1 X_i + \beta_2 Y_i + e_i$$

which states that the elevation Z at some point i is equal to a constant, α, plus contributions from the slope in the X-direction and the slope in the Y-direction, plus a deviation, e_i. These relationships can be readily seen

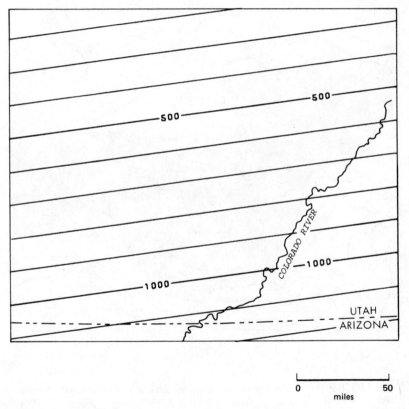

FIGURE 4.8.   First-degree trend surface of thickness of Cedar Mesa Sandstone. Contours in feet.

in Figure 4.6 and in cross-sectional form in Figures 3.6 and 3.7. The coefficients of the equation are found by methods which ensure that the sum of the squared deviations from the fitted line is a minimum in the direction of the ordinate or Z-axis. This is equivalent to saying that the variance about the trend is as small as possible. In the example shown in Figure 4.6, the equation is

$$Z = -1329 - 1.5X + 16.6Y$$

This means that at the origin, the trend surface is 1329 feet below sea level

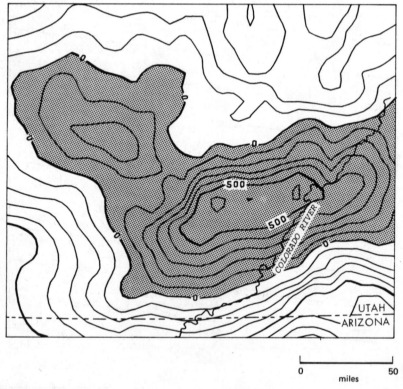

FIGURE 4.9. Residuals from first-degree trend of thickness of Cedar Mesa Sandstone, obtained by subtracting trend in Figure 4.8 from isopach map in Figure 4.7. Positive residuals are shaded. Contour interval is 100 feet.

and slopes to the west at 1.5 feet per mile and to the south at 16.6 feet per mile. The first-degree trend accounts for 41 percent of the variation in the structural surface.

In an exploration context the trend surface is usually taken to represent a regional component or regional "trend," and the deviations are regarded as "local" features. Figure 4.7 is an isopach map of the Permian Cedar Mesa Sandstone in the Four Corners area (Irwin, 1971). A first-degree trend surface (Figure 4.8) has the form of a gently dipping plane; the thick central area is isolated as a positive residual (Figure 4.9) from this surface.

However, it is possible (and in this example, likely) that a simple linear

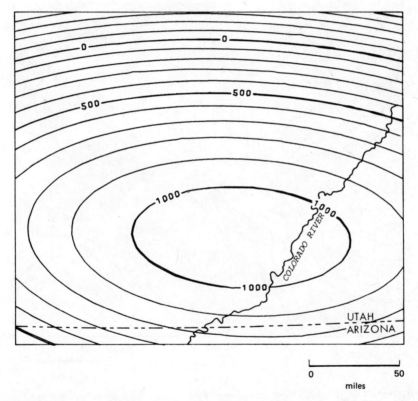

FIGURE 4.10. Second-degree trend surface fitted to thickness data for Cedar Mesa Sandstone (Permian) in Four Corners area. Contours in feet.

trend will not adequately represent the "regional" component. In such a case, the residuals will themselves be of regional scale and a satisfactory separation of "local" features will not be achieved. If polynomials are used, the trend surface equation can readily be expanded by adding terms which are powers of the geographic coordinates of the sample points. If expanded to the second degree, the equation becomes

$$Z_i = \alpha + \beta_1 X_i + \beta_2 Y_i + \beta_3 X_i^2 + \beta_4 Y_i^2 + \beta_5 X_i Y_i + e_i$$

Each term added to the equation gives the trend surface one additional

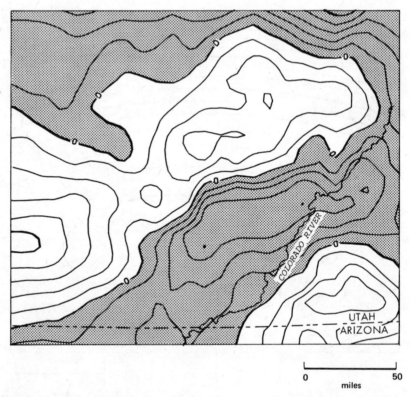

FIGURE 4.11. Residuals from second-degree trend, calculated by subtracting trend surface in Figure 4.10 from isopach map in Figure 4.7. Positive residuals are shaded. Contour interval is 100 feet.

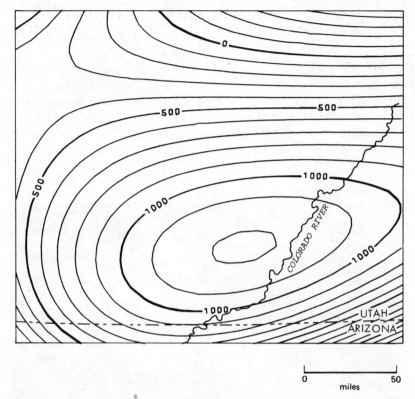

0   50

miles

FIGURE 4.12.   Third-degree trend surface fitted to Cedar Mesa Sandstone data from Figure
4.7. Contours in feet.

degree of flexibility. The surface may now bend in either the X-direction,
the Y-direction, or in both. Because the surface is more flexible, it will
conform more closely to the data. That is, the trend will be a better
approximation of the form of the surface, and magnitudes of residuals will
be smaller. This can be seen in the cross-sections of Figure 3.7, and in the
second-degree trend and residual maps of the Cedar Mesa isopach (Figures 4.10 and 4.11).

If a greater reduction in the size of residuals is desirable, the trend
equation can be expanded to still higher powers. A third-degree equation
is

$$Z_i = \alpha + \beta_1 X_i + \beta_2 Y_i + \beta_3 X_i^2 + \beta_4 Y_i^2 + \beta_5 X_i Y_i$$

$$+ \beta_6 X_i^3 + \beta_7 Y_i^3 + \beta_8 X_i^2 Y_i + \beta_9 X_i Y_i^2 + e_i$$

This surface may reverse its dip twice in each geographic direction. Consequently, it is extremely flexible and may conform closely to the general form of the data. Again, Figure 3.7 shows the reduction in magnitude of residuals in two dimensions, and Figures 4.12 and 4.13 show that a third-degree trend surface leaves small residuals from the Cedar Mesa isopach.

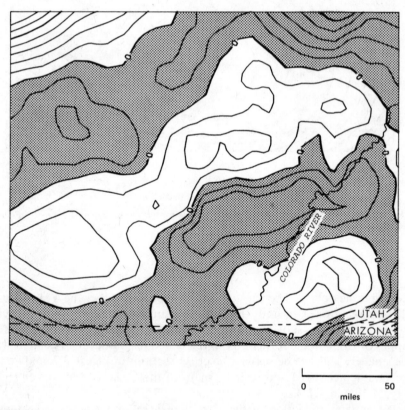

0   50
miles

FIGURE 4.13. Residuals from third-degree trend, calculated by subtracting trend surface in Figure 4.12 from isopach map in Figure 4.7. Positive residuals are shaded. Contour interval is 100 feet.

In theory, the trend equation may be expanded until the surface becomes so flexible and complex in shape that it passes exactly through every data point and the residuals vanish. However, there would be no purpose to such an exercise, because the objective of trend surface analysis is to isolate local features from the broader pattern of geologic variation. Choosing the proper degree for the trend equation requires experimentation and empirical analysis. Unlike filtering, there is no way to determine the size of features that will be isolated prior to fitting the trend. However, the most appropriate trend surface can be selected *a posteriori* on the basis of its ability to isolate productive features, as we see later in a case study.

CHAPTER 5

Probability Analysis of Computer-Contoured Structure

ESTIMATING THE ERROR IN CONTOUR MAPS

For the purpose of preparing a map, a geologic variable can be known with certainty only at the control points, where it is observed or measured directly. Although there may be errors involved in measuring the variables at the control points, these can be considered distinct from errors resulting from interpolation in the mapping process. Thus, away from control points within the mapped area, the geologic variable is estimated either by a contouring program or manually. An estimate is simply an inspired guess, and like most guesses, it is usually in error. We would hope, however, that the estimates are not seriously incorrect and that the magnitudes of these errors over most of the map are within acceptable limits.

In the absence of complete information about the nature of the variable being mapped, errors are inevitable. These errors will be greater for a geologic variable that changes rapidly within small areas than for a variable that changes slowly and consistently. A map constructed from a dense collection of control points will have less error than one made from only a few points. Similarly, a map made from control points spaced on a regular, uniform pattern such as a square grid will be more consistently reliable than a map constructed from irregularly spaced control points. Both conditions reflect the interrelationship between spatial autocorrelation of the surface, the spacing between control points, and the likely error at interpolated points.

These interrelationships are formalized in regionalized variable theory, a branch of statistics devised initially for the analysis of mining problems (Matheron, 1965). Under this statistical theory, geologic variables are considered to be composed of two components, a gentle regional trend, or "drift," and autocorrelated residuals or deviations from the drift. Once the drift has been removed (usually by subtracting a series of local trend surfaces or moving averages), the degrees of similarity between points is given by the semivariance. The semivariance is the variance of the differences between all possible pairs of points on the map that are the same distance apart and is closely related to the autocorrelation. A plot of semivariance versus distance is called a semivariogram (Figure 5.1) and

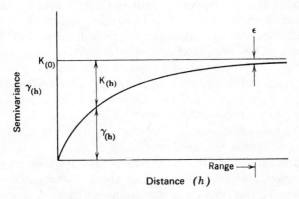

FIGURE 5.1. Semivariogram, or plot of semivariance (γ) versus distance. Range, indicated by arrow, is distance at which semivariance is essentially equal to variance of surface.

rises from zero at zero distance (two measurements made at the same point will be identical) to a maximum value equal to the total variance around the drift. The distance at which the semivariance becomes a maximum is called the range, and it is also the distance at which the autocorrelation of the mapped surface will drop to zero. The semivariogram expresses the increasing uncertainty associated with estimates made at progressively greater distances from known control points.

Universal Kriging

Once the form of the semivariogram has been experimentally determined, it may be used in a surface estimation procedure called universal kriging

(Matheron, 1969; Huijbregts and Matheron, 1971; Huijbregts, 1975). This is a method of contouring in which maximum use is made of the autocorrelation in the surface being mapped. The resulting contour map is optimal in the sense that it will have the smallest estimation errors possible with any estimating procedure, provided the drift has been properly removed. In addition, kriging also yields a measure, in the form of a map, of the possible errors at all estimated points. This is a map of the "plus-or-minus" bounds on the mapped surface at every point within the map area. Universal kriging has been widely applied in mining and mineral exploration (David, 1972; Huijbregts and Journel, 1972; Huijbregts and Segovia, 1973), but there are few published examples of its application to petroleum exploration. Olea (1974), however, demonstrated the use of universal kriging for subsurface structural mapping.

Olea and Davis (1977) also applied universal kriging to map subsurface structure in the Magellan basin of southern Chile (Figure 5.2). Oil production in the area comes from combined structural and stratigraphic traps in sandstone members of the Cretaceous Springhill Formation, a sequence of alternating sandstones and shales overlain by impermeable deep marine shales. Traps consist of penecontemporaneous anticlinal structures in the Springhill Formation. However, reservoir sands may be absent due to erosion or nondeposition over the crests of the anticlines, resulting in complex annular reservoirs around barren cores.

One of the most prolific areas of production in the Magellan basin is the Cullen field, which has produced over 30 million barrels of oil to date. The 163 wells drilled in the vicinity of the field provide basic information about its configuration. However, Olea utilized another source of data to determine the semivariogram of the structural horizon. A marine seismic survey in the Straits of Magellan immediately north of the Cullen area provided 2000 equally spaced observations from which the semivariogram could be calculated. The coefficients of the semivariogram were then used to contour the geologic structure in the Cullen area (Figure 5.3) and to estimate the associated error map (Figure 5.4). Because of the discontinuous nature of the Springhill Formation, structure was mapped on the top of the underlying Jurassic Tobifera Series, a thick sequence of tuffs, breccias, and ignimbrites.

The error map can be thought of as an isopach map, defining an interval between the likely upper and lower bounds within which the true structural elevation of the Tobifera lies. If we assume that the errors in estimation between control points are normally distributed, the probability is almost 70 percent that the true surface lies within plus or minus the interval shown on Figure 5.4, and the probability is 95 percent that the true surface elevations will be within double this interval.

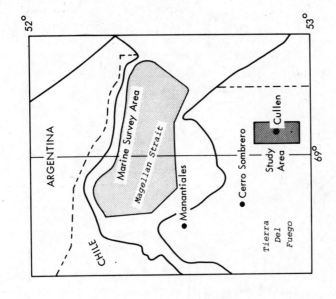

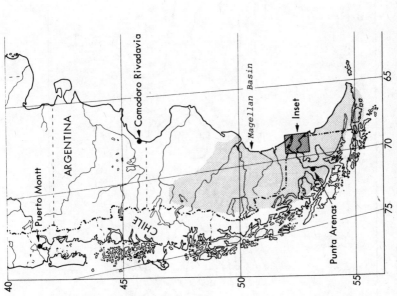

FIGURE 5.2. Index map of Magellan basin area mapped by Olea and Davis (1977). Shaded portion indicates Magellan basin. Inset shows location of marine seismic survey area in Straits of Magellan and mapped area around Cullen field.

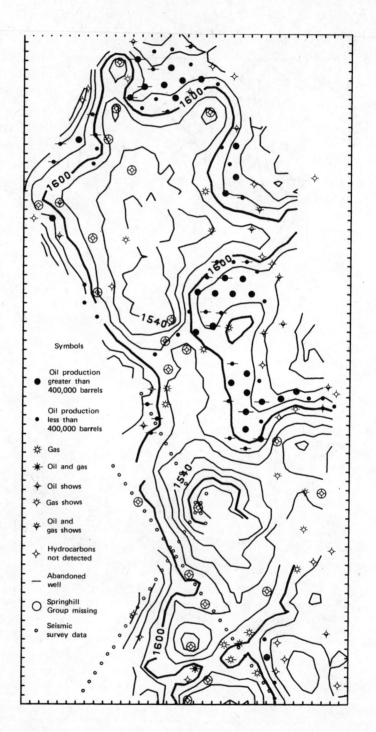

Symbols

● Oil production greater than 400,000 barrels

• Oil production less than 400,000 barrels

☼ Gas

✳ Oil and gas

✦ Oil shows

✧ Gas shows

✴ Oil and gas shows

✧ Hydrocarbons not detected

— Abandoned well

◯ Springhill Group missing

∘ Seismic survey data

FIGURE 5.3. Structure contour map of top of Jurassic Tobifera Series in the Cullen area of Magellan basin, Chile, prepared by universal kriging. Contours in meters below sea level. Contour interval is 20 meters. Horizontal bar through well symbol indicates abandoned well. Circle around well symbol indicates Springhill Group missing. Contours not drawn in areas of insufficient control. After Olea (1975).

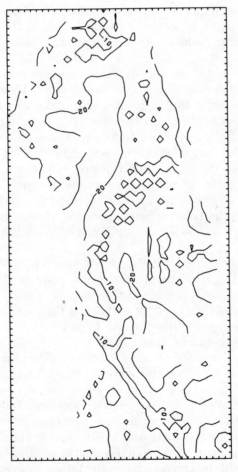

FIGURE 5.4. Error map of structure contour map shown in Figure 5.3. Contours in meters correspond to ±one standard deviation. After Olea (1975).

The kriged map provides valuable geological information about the structure of the Cullen area, and the error map provides information about the reliability of these structural estimations. Certain, probable, and fictitious structures can be differentiated in an objective manner. The error map and the semivariogram can also be used to determine where more information is necessary in order to refine the map and to estimate the number of additional observation samples that will be needed. In this way, regionalized variable theory provides criteria by which future sampling can be planned to achieve specified levels of reliability. The estimation error can be reduced to a prespecified limit by taking more samples in these critical areas.

Of even greater exploration interest is the drift of the mapped geologic surface, which is the expected or average value of the surface within a local area or neighborhood. The concept of drift, similar to the more familiar "trend," was introduced by Matheron (1971) who felt that the word trend had been misapplied. The purpose of drift estimation is to provide a map that emphasizes local structures which may be especially prospective. The drift itself is not the most useful way to visualize such areas and a map of the residuals from the drift may be more revealing, just as trend surface residuals are useful for exploration. However, for a given geologic variable, the neighborhood size and drift equation must be adjusted to achieve the appropriate isolation of features. The problem is analogous to selection of an appropriate order for trend surface analysis, except that more parameters must be specified. Figure 5.5 shows a drift map of the top of the Tobifera Series using a neighborhood having a radius of 2.1 kilometers and a first-degree polynomial for the drift within the neighborhood.

In the Cullen field, the suitable drift will yield a map of residuals which reflect sedimentary environments when the Springhill Formation was deposited and which can be used to forecast oil production. Figure 5.6 is the map of residuals obtained by subtracting the estimates for the top of the Tobifera (Figure 5.3) from the drift (Figure 5.5). The contour lines on the drift map appear to be very complex. However, this apparent complexity is solely a function of the low relief of the surface. The residuals are subdivided into three classes, two of which are shaded in the illustration.

Wells in the Cullen field have been grouped into four categories: (1) high oil production—with a cumulative production of over 400,000 barrels, and recently drilled wells with above-average initial production; (2) oil or gas—remaining wells with commercial production of hydrocarbons; (3) dry—wells without commercial production of hydrocarbons from the Springhill Formation; (4) wells in which the Springhill is missing. The

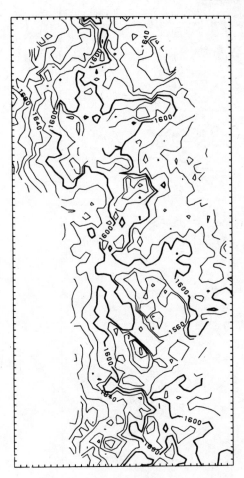

FIGURE 5.5. First-degree drift of top of Tobifera Series, Cullen area, made using 2.1-kilometer neighborhood radius. Contours in meters below sea level. Contour interval is 20 meters. After Olea (1975).

negative drift residuals shown in Figure 5.6 reflect the large embayments formed on the northeastern portion of the Cullen field during marine transgression at the end of the Jurassic. Sand deposition occurred only in bays open to the north or east and did not occur on the promontories between bays. Therefore, the distribution of potential reservoir facies in the Springhill Formation is closely related to structure of the underlying unit. In the Cullen field, oil production and drift residual magnitude are highly correlated, as shown in Figure 5.7.

Of all wells drilled where the Springhill Formation is missing, 86 percent are in areas where the drift residual is above 30 meters. On the other hand, 80 percent of the wells in areas with negative residuals are in

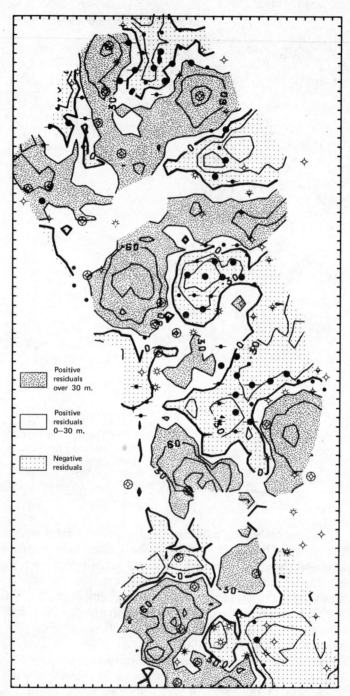

FIGURE 5.6. Residuals from first-degree drift, obtained by subtracting structure map from drift map. Contour interval is 30 meters. After Olea (1975).

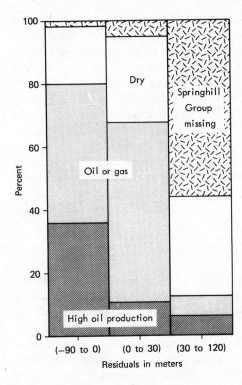

FIGURE 5.7.   Proportions of wells in four production categories in Cullen field with respect to three classes of residual magnitudes mapped in Figure 5.6 (Olea, 1975).

productive categories, and the Springhill Formation is missing in only 2 percent of these wells. The percentage of high-production wells is more than six times greater in negative residual areas than in areas of residuals over 30 meters. The percentage of wells in all productive categories is almost seven times greater in the negative residual areas than in areas with residuals over 30 meters.

The areas having residuals of intermediate magnitude are also intermediate in production. The percentage of wells having high productivity is less than one-third of that in the negative residual areas; only 67 percent of the wells are commercial producers compared to 80 percent in the negative residual areas. The percentage of wells in which the Springhill Formation is absent is twice of that in the negative residual areas.

Wells which penetrate the Springhill Formation but do not produce in commercial quantities are almost equally distributed in the three categories of residuals. However, in areas with residuals larger than 30 meters, the wells tend to be dry because the sediments have low permeability, whereas in the negative residual areas the sandstone usually has good permeability, but water-saturated porosity.

Although universal kriging and drift analysis have extremely attractive properties, they unfortunately present formidable computational problems. Every estimated point on a map requires the solution of a set of simultaneous equations involving all control points within a distance equal to the range around the location of the estimated point. As a consequence, computer charges for universal kriging may run 3 to 10 times those for conventional automatic contouring (Davis, 1976). Perhaps more seriously, kriging must be preceded by a manual step called "structural analysis" which essentially involves determination of the appropriate order of the drift and the form of the semivariogram. This must be done by trial and error and requires subjective but experienced decisions. Unless the structural analysis is performed correctly, the kriging process will not be optimal and the method will have no advantages over conventional contouring.

Empirical Estimates of Error

Estimation errors in automatic contouring can be assessed empirically, by using a subset of the available data to predict values at "blind" locations where the true values are already known. This technique is especially applicable in studies of the type considered in this book, because it is closely related to the methods used to find conditional relationships between perceived geology and oil occurrence. For example, maps may be made of the geologic structure in a particular area as it would have been perceived at an early stage in the exploration history of that area. This involves a prediction of the structural elevation at locations where wells were subsequently drilled. The actual elevations at these wells thus provide an empirical measure of the error in the initial contour map.

Figure 5.8 is a contour map of structure on top of the Pennsylvanian Lansing–Kansas City Group as it would have been mapped in Graham County, Kansas in 1952, using the 352 wells then available. This same area is used in the next section to illustrate the calculation of success probabilities. Except at the well locations indicated by crosses, structural elevations have been estimated by the SURFACE II contouring system. Errors in the estimates can be determined by comparing this map with Figure 5.9, which is based on 2758 wells that had been drilled in the same area prior to 1974. Differences between these two maps provide 2406 measurements of estimation error in the 1952 map. That is, for each well drilled in the period 1953 to 1973, there is a true elevation Z_i and an elevation Z_i' estimated from pre-1953 well data. The difference is the estimation error $\epsilon_i = Z_i - Z_i'$.

Intuitively, it seems reasonable that the magnitude of estimation errors should be less for locations near control points and greater for more

FIGURE 5.8. Structure of top of Lansing-Kansas City Group as perceived in 1952, when 352 wells had been drilled in Graham County area. Producing wells are indicated by black dots, dry wells by crosses. Contour interval is 20 feet. Areas below −1300 feet are shaded. Tick marks along borders are at 1-mile intervals (Hambleton, Davis, and Doveton, 1975).

distant locations. If this assumption is true, it should be possible to express error magnitude as a function of distance from control. This possibility may be assessed by calculating the distance from the location of each post-1952 well to the nearest well drilled in 1952 or earlier. (This is performed automatically as part of the contouring process, as the program must calculate the distance to the control points to determine their weightings.)

Each estimation error ϵ_i is now associated with a measurement of the distance d_i to the nearest control point. In this example, d_i ranges from

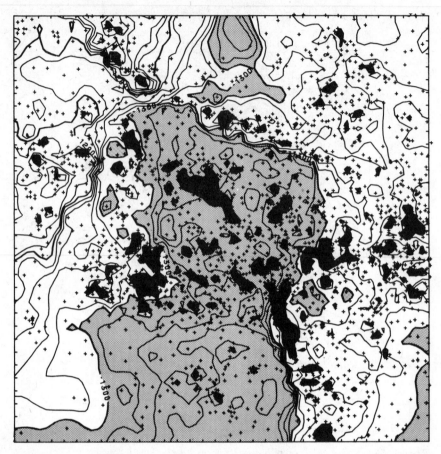

FIGURE 5.9. Perception of structure of top of Lansing-Kansas City Group in Graham County area in 1974. Area contains 2758 wells. Map conventions are same as in Figure 5.8 (Hambleton, Davis, and Doveton, 1975).

less than one-quarter mile to over 6 miles; most of these distances, however, are less than 2 miles. In other words, since 1952 it has become difficult to locate a wildcat well in the study area that is not within 2 miles of a preexisting well location. The distances are grouped into one-quarter-mile intervals. For each distance interval, the root mean squared (RMS) error is calculated. This is the square root of the mean of the squared errors and is a measure of the average deviation of the estimated Z_i' values from the true Z_i values. The RMS error is equivalent to the standard deviation of the errors within an interval, but is not corrected for their mean. This is because the estimated surface may be biased (i.e., consistently too high or too low) and this source of error will not be reflected if the mean (which would also be biased) is subtracted.

Figure 5.10 is a plot of RMS error for each quarter-mile interval and is based on data contained in Table 5.1. Each of the first eight intervals include large numbers of observations, so the RMS values calculated for these distances can be regarded as highly reliable. The longer distance intervals contain fewer observations and hence are more erratic. It is especially significant that the most reliable values lie almost exactly on a straight line. This line has been plotted on Figure 5.10 and provides a conservative measure of the increase in estimation error as a function of distance away from the nearest control point. For this example, the function is approximately 14.6 feet of vertical error per mile of distance from the nearest control point.

Figure 5.11 shows histograms of the error for successive quarter-mile distances up to 1¼ miles from the nearest control well. Although it is necessary to group the errors into histogram categories in order to calculate relative frequencies, the errors can be thought of as forming a continuous distribution. Figure 5.12 is equivalent to Figure 5.11, but the

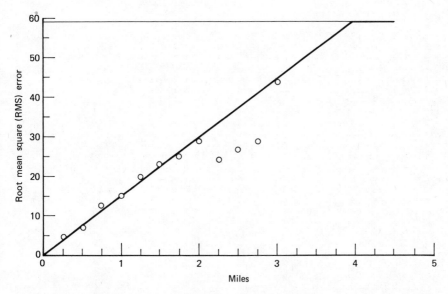

FIGURE 5.10.   Root mean square (RMS) error in structural surface estimated from 1952 well data from Graham County training area in northwest central Kansas, expressed as function of distance to nearest control well. Upper limit is standard deviation of structural elevations in training area (Hambleton, Davis, and Doveton, 1975).

TABLE 5.1. Root mean squared error in contoured surface estimates at 2406 locations in Graham County, Kansas based on differences between a structural map based on pre-1952 data, and a second map based on pre-1974 data. RMS errors are tabulated according to distance to nearest pre-1952 well.

Distance (Miles)	RMS Error (Feet)
0.25	4.5
0.50	7.2
0.75	12.6
1.00	14.7
1.25	20.0
1.50	22.6
1.75	24.9
2.00	28.7
2.25	24.2
2.50	26.6
2.75	28.5
3.00	43.9

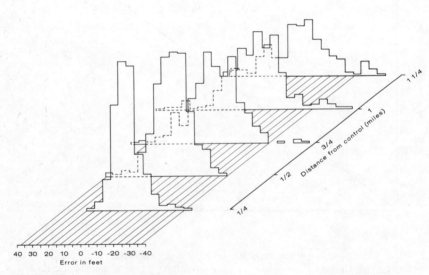

FIGURE 5.11. Histograms of estimation errors plotted for successive quarter-mile distances from nearest preexisting well.

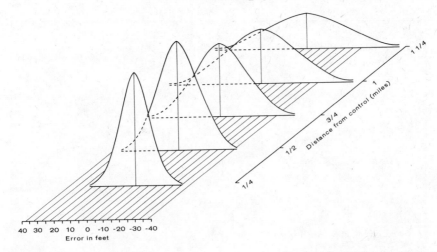

40 30 20 10 0 -10 -20 -30 -40
Error in feet

FIGURE 5.12. Normal error curves having same means and variances as histograms shown in Figure 5.11, plotted for successive quarter-mile distances from nearest preexisting well.

histograms have been replaced with normal curves having the same means and variances as the error histograms. Moments of the data indicate the errors are more tightly clustered about the true values than would be expected with a normal curve. Therefore, if we assume the errors are normally distributed for the sake of computational convenience, we are being conservative. In fact, the observed estimation errors are less extreme than predicted by a normal distribution.

Every point on a map can be characterized by its distance from the nearest of a set of control points, since this distance must be computed for every point in the map array (Sampson, 1975a). These maps of distance can be converted easily to maps of expected estimation error (Figure 5.13) by multiplying the distance map values by the corresponding error function.

If the estimation errors are assumed to be normally distributed, the RMS error map can be interpreted as a "confidence envelope" around the estimated surface. The error map can be added to and subtracted from the estimated surface to form a "lowest likely" and "highest likely" structural surface (Figures 5.14 and 5.15). By assuming that the errors are normally distributed, we can state that approximately two-thirds of the true surface elevations will be contained between these two limits. In a complementary sense, we may use the error map in conjunction with the structural contour map to obtain a point estimate of the structural elevation at an undrilled location, and a "plus-or-minus" statement of the possible error or uncertainty in this estimated elevation.

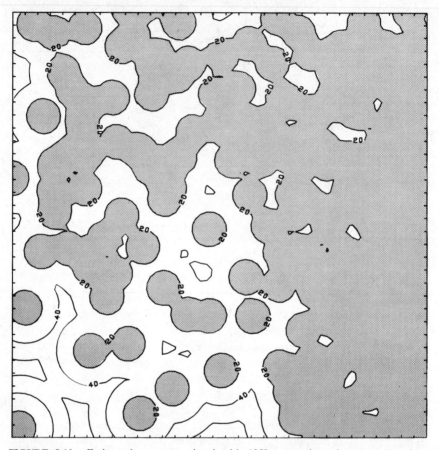

FIGURE 5.13. Estimated error associated with 1952 perception of structural surface shown in Figure 5.8. Contour interval is ±20 feet error around estimated structural surface elevations.

The empirical relationship between estimated error ϵ_i and distance from control d_i has the form of a sloping straight line (Figure 5.10). Such a relationship implies that the uncertainty or estimation error increases without bounds as the distance from control becomes greater. However, it is obvious that this cannot be the case. The maximum error that is likely to be made is fixed by the total variation in elevation of the structural surface within the map area, as this describes deviations that would exist around an estimated "surface" which was no more than a flat, featureless plane at the average elevation of the structural horizon. The error func-

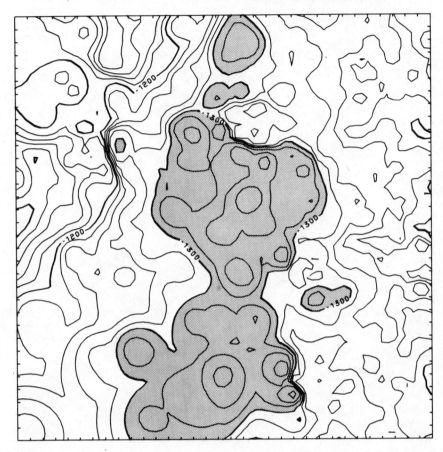

FIGURE 5.14. "Highest likely" structure map, based on 1952 structure map (Figure 5.8) and error map (Figure 5.13). Only 16 percent of estimated points are likely to be structurally higher than elevation given.

tion can be expected to increase only to an upper limit equal to the standard deviation of elevations in the structural surface. For the Lansing-Kansas City surface in Graham County, this limit is 54.8 feet as shown on Figure 5.10. Therefore, the error map in Figure 5.13 is unduly pessimistic for locations more than about 4 miles from the nearest control point.

If an isopach is made between two structural surfaces, the maximum possible uncertainty in thickness is equal to the sum of the uncertainties associated with the two surfaces. In practice the uncertainty usually is

FIGURE 5.15. "Lowest likely" structure map, based on 1952 structure map (Figure 5.8) and error map (Figure 5.13). Only 16 percent of estimated points are likely to be structurally lower than elevation given. Two-thirds of true structural elevations will lie between limits given by this map and preceding map (Figure 5.14).

less, because the two structural surfaces may be highly correlated. For linear transformations of a surface, such as trend surfaces and their residuals, the error function remains the same as for the original surface. Error analysis can be a useful aid in exploration where prospects are based on local highs or trend residuals, because it allows a probabilistic assessment of the actual magnitude (or even the existence) of an exploration target.

MAPPING OF TRAINING VERSUS TARGET AREAS IN PROBABILISTIC FORECASTING

At this point it is appropriate to consider the use of machine-contoured maps in statistical forecasting of oil occurrence. The general approach is to obtain outcome probability estimates in relatively mature oil-field areas and, in turn, to use these estimates in immature areas that are now undergoing active exploration. The mature area may be considered a training area, to be analyzed by reexperiencing the drilling history to yield outcome probabilities that are based on a succession of trials. Assuming that oil has been formed and trapped under similar conditions in the training and exploratory areas, the outcome probabilities obtained in the training area can be employed in the exploratory area.

Assuming that oil occurs under more-or-less equivalent conditions in two different areas may be unwarranted in some circumstances. However, assumption of a degree of geologic similarity of oil occurrence is implicit in any geologically guided exploration program, whether a probabilistic approach is employed or not.

As part of the Kansas Geological Survey's KOX project, a large area in northwestern Kansas (Figure 5.16) and a smaller extension in adjacent Nebraska were analyzed to determine the relation between perceived

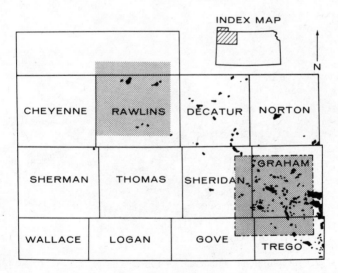

FIGURE 5.16.   Index map of northwestern Kansas region, which contains Rawlins County target exploratory area and Graham County training area.

geologic structure and oil occurrence. A 30 × 30 mile square subarea was selected as a training site in a densely drilled part at the northwest end of Central Kansas Uplift. All oil production in the training area comes from limestone reservoirs in the Lansing-Kansas City Group, of Pennsylvanian age. The remainder of the area shown in Figure 5.16 can be considered as undergoing active exploration. Attention was concentrated on Rawlins County, which has the same dimensions as the training area. Discoveries in Rawlins County also have been confined to the Lansing-Kansas City Group, and the general mode of oil occurrence appears comparable to that in the training area.

Oil occurrence in the Graham County training area shows a strong relationship to structural configuration of the producing horizon. Figure 5.17 is a block diagram of the top of the Lansing-Kansas City Group in area, as interpolated from 2758 wells that had been drilled before early 1974. The major structural feature consists of a broad central basin whose northern margin roughly coincides with the flank of the Central Kansas Uplift. The basin is bounded to the east and west by positive structural blocks and probably results from late downwarping movements along persistent lines of weakness associated with the uplift. The distribution pattern of existing oil fields shows a strong concordance with local structural highs (Figure 5.9), implying that Lansing structure is highly diagnostic in the prediction of field locations.

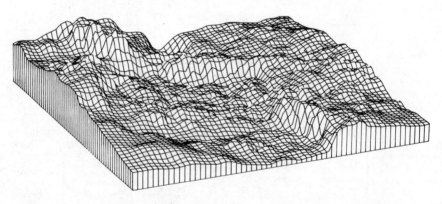

FIGURE 5.17. Structural block diagram of top of Lansing Group (Pennsylvanian) in Graham County training area, based on 2758 wells available in 1974. View looking north.

Historical Analysis of Training Area

Well control in the training area is sufficiently dense that the interpolated structural surface is a close approach to the "real" Lansing surface if this could be observed with perfect resolution. However, this was not the case in the past when only limited information was available. The perception of any geological surface, both at regional and local scales, undergoes a dynamic evolution with time as the network of well control is expanded across the area. Figures 5.8 and 5.9, for example, show the Lansing surface as it would have been perceived by interpolation from the available well control in the years 1952 and 1974. In each of these years, the perceived structural surface would have been the major guide for exploration decisions executed in the following year.

The error function described earlier may be used to map possible error in structural maps of the training area, as has been done in Figure 5.13. Because of the increase in available well control, the uncertainty associated with perception of the Lansing surface diminishes through the historical sequence. In 1937, the uncertainty attached to any prediction of subsurface configuration was very high. At the present time, there is a uniformly low degree of uncertainty except in the extreme southwest corner of the training area.

Application to the Target Area

Approximately 349 wells have been drilled in the Rawlins County target area and it is therefore at a similar stage of exploration as the Graham County training area in 1952, when 352 wells had been drilled. The two areas can be considered to be in parallel time streams of exploration activity, with the status of the target area in 1974 broadly equivalent to that of the training area in 1952. The Lansing surface in the target area (Figure 5.18) shows a diffuse north-south ridge across the central part of the area, connecting positive structural features in the northeast and southwest. If local surface variability in the Lansing is effectively equivalent in training and target areas, then hindsight analysis of the training area provides a viable tool for predictions in the target area. Application of the uncertainty function to the target area control network results in an error map (Figure 5.19) that is a corollary to the interpolated surface estimation map in Figure 5.13. The two maps are a necessary preliminary step in the analysis of probabilistic relationships between Lansing structure and oil-field locations.

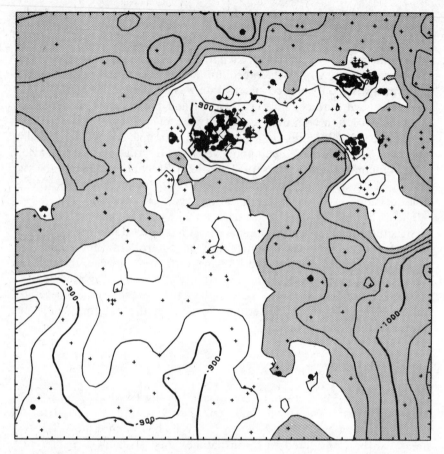

FIGURE 5.18. Structure of top of Lansing Group in Rawlins County target area as perceived in 1974. Area contains 358 wells.

Relationships Between Structure and Oil-Field Locations

Explorationists characterize structure, not in absolute terms of feet above or below sea level, but in relative terms such as anticline or syncline. Unfortunately, it is difficult to quantify these expressions of local shape, even though they are the elements of structure that control the entrapment of petroleum. The procedures that involve qualitatively defined local structure within cells described in Chapter 2 are subjective assessments and hence liable to bias. The measures described in Chapter 4 include quantitative expressions of the rate of curvature of a surface

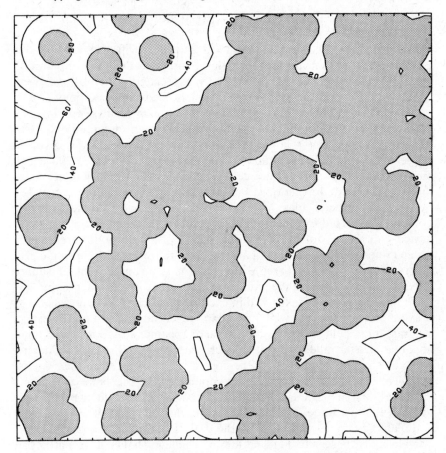

FIGURE 5.19. Error map for perceived structure in Rawlins County target area.

around the point being classified, and also encompass map filtering, which passes only structures of specified dimensions, and separation of structure into "local" and "regional" components by trend surface analysis. Because trend surface analysis is computationally among the most straightforward of these methods, it was used to provide the measure of local structure. Also, trend surface analysis is routinely used by explorationists in northwestern Kansas and thus has the added advantage of industry acceptability. A polynomial trend surface tends to conform to the regional structural grain, leaving smaller features as residuals or deviations. At any point, either a well location or prospect site, local structure may be measured as the magnitude of the deviation from a trend.

The probability that a randomly located wildcat will discover a field is proportional to the total area of fields relative to the total exploration area. If oil occurrence is preferentially associated with positive trend residuals, the proportion of positive residuals occupied by oil fields will be greater than the proportions elsewhere in the map area. A contingency table showing the conditional relationship between magnitude of trend surface residuals and relative area underlain by production is given in Table 5.2. Residual magnitudes were taken from Figure 5.20, which shows fourth-degree trend surface residuals for the Graham County training area, calculated using only well information available through 1952.

Areas within each class of residual magnitude were found from the computer grid used to generate Figure 5.20. First, values in the grid were sorted into the six intervals of Table 5.2. As the grid nodes are equally spaced one-quarter of a mile apart, the area represented by a single grid cell is 40 acres. The total area within any interval can be found in square miles simply by dividing the number of grid nodes in that interval by 16.

To find the areas of fields within each residual class, it is necessary to assume that a producing well taps a fixed reservoir area. In northwestern Kansas, most field wells are on a 40-acre spacing, so we may assign an area one-quarter mile on a side to each producing well. Every producing well can then be equated, with negligible error, to a cell in the residual map. This is done by generating a computer grid of the same dimensions as the residual map, in which the grid node nearest a producing well is set to 1 and all other nodes are set to a blank or default value. If this grid of 1s and blanks is cross-multiplied by the grid of residuals, the product grid contains only the residual magnitudes of producing fields. This matrix can also be sorted into the six intervals of Table 5.2. Finally, the proportion of

TABLE 5.2.   Percentage of areas occupied by fourth-order trend residuals that are underlain by known petroleum reservoirs in Graham County, Kansas. Residuals are based on Lansing-Kansas City structure.

Residual Magnitudes	Percentage of Area
$-\infty$ to -40	1.63
-40 to -20	3.74
-20 to 0	7.26
0 to 20	15.36
20 to 40	20.15
40 to ∞	39.26

FIGURE 5.20.   Fourth-degree trend surface residuals calculated from well control available in Graham County training area in 1952. Positive residuals are shaded. Contour interval is 20 feet.

each residual class underlain by production can be found by comparison of these values with the total areas of the residuals. The resulting proportions give the probability of successfully hitting production if a wildcat is drilled into a trend surface residual of specified magnitude.

Note that although the residuals are based on the data available only through 1952, the field areas are based on all available data. Thus, the probabilities given in Table 5.2 relate to the likelihood that a structure, defined on the basis of limited prior information, will ultimately produce. With additional post-1952 information, the pattern of residuals (and hence the probabilities) may change. However, the areas of production will not,

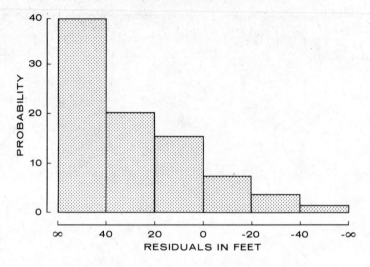

FIGURE 5.21. Conditional probability distribution showing likelihood of success in drilling trend surface residuals of perceived magnitude, based on Graham County data available through 1952. Taken from Table 5.2.

unless significant new fields are found in the training area (an unlikely circumstance in view of the 1974 drilling density).

The conditional probability distribution in Figure 5.21 suggests that trend surface residuals could have been used directly to guide petroleum exploration in Graham County. However, the perception of residuals is subject to the same uncertainty as perception of the structure itself, because the residual map is a linear transformation of the structural surface. Therefore, the probability of success at any prospect site must be adjusted to take into account the uncertainty in residual magnitude, which reflects the relative density of well control in the local neighborhood, in exactly the same manner as the structural map itself was adjusted. At any point in the training area, the perceived trend residual may be regarded as a point estimate of the true residual and the uncertainty as the standard error of this estimate. If the errors are assumed to be approximately normally distributed, an adjusted probability of success can be calculated. This is done by multiplying the probability of success in a class of residual magnitudes by the probability that the residual will actually fall into that class. Summing these products across all classes yields the adjusted probability of success for that location (Figure 5.22).

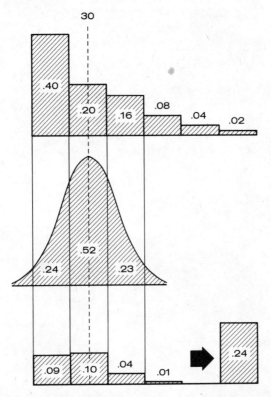

FIGURE 5.22. Cross-multiplication of conditional probability distribution by error distribution and summation yields cumulative probability of success. Example shown is for perceived 30-foot positive residual located one-half mile from nearest control.

Figure 5.23 is a map of the probability of success for drilling ventures in Graham County, as it would have been perceived at the end of 1952. Areas predicted to have a producing/dry hole ratio higher than the ambient or average level of 0.10 are shaded. This illustration should be compared with Figure 5.9, which shows oil fields subsequently discovered after the time for which the probability map was calculated. The probability map successfully predicts fields that were later found in the northwestern corner and southwest of the center of the county. These discoveries are especially encouraging because no production had been found in either area as of 1952. The map also predicted the multimillion-barrel Hoof field southeast of the center of the county, although nearby production was already leading explorationists toward its discovery. The

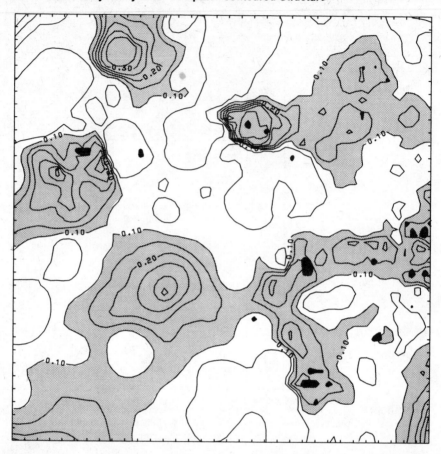

FIGURE 5.23.   Probability map for Graham County training area, based on structure as perceived in 1952. Contour interval is 0.05 probability of discovery of oil. Areas having probabilities higher than ambient level of 0.10 are shaded. Fields known to exist in training area prior to 1952 are indicated in black; compare with Figure 5.9 to determine association of post-1952 fields with areas of high probability.

probability map failed to predict the discovery of the complex of fields in the structurally low region in the center of the county. This reflects in part the low drilling density available in this area in 1952, which resulted in relatively high uncertainty about the local structural configuration. There is also evidence that these reservoirs are not structurally controlled, but are instead stratigraphic or combination traps. As the probability map is based solely on structural configuration, failure of the method to predict these fields is not surprising.

Because the probability map is based on post-1952 well outcomes, it might be expected that each category would contain exactly the number of outcomes as in the original contingency table. This, however, is not the case, because of the uncertainty of perception. Probabilities in rank wildcat areas will be adjusted up or down toward the regional or ambient probability. This probability is the "level of indifference," or the success ratio that an operator might anticipate if he were to drill at random making no attempt to select promising sites or to avoid unfavorable locations. It

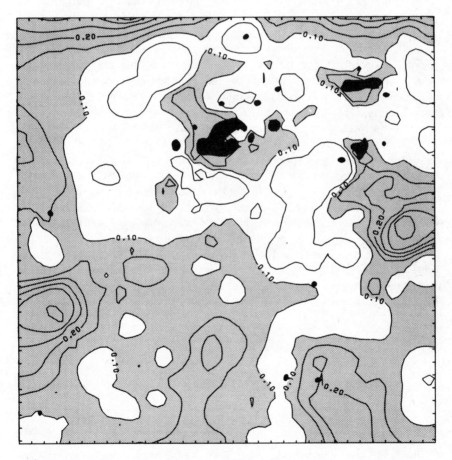

FIGURE 5.24. Probability map of Rawlins County area based on structural data available at end of 1974. Areas having probabilities higher than ambient level of 0.10 are shaded.

also should be emphasized that as drilling continues in an area, the probability map will change, reflecting not only an increase in perception but also a decline in the ambient probability as the potential targets are successively tested. However, updated probability maps will continue to indicate the relative merits of potential sites within an area, even though these might become quite discouraging.

Figure 5.24 is a probability map for the Rawlins County test area, calculated using all information available at the time the map was created (1974). It is an estimate of the success that will be achieved by explorationists in the area in the future. Probabilities are based on the conditional relationship between structural geology and the discovery of oil established in the Graham County training area. These probabilities are adjusted for the uncertainty in perception of structure resulting from the limited number of control wells available. The function relating uncertainty in perception to distance from nearest available control has also been derived from the training area. Therefore, this map contains an additional source of uncertainty which depends on the true degree of similarity in structural styles and oil occurrence between the training and target areas. Of course, such uncertainty is always present whenever an area is interpreted in light of experiences in other areas.

This particular example is based on a two-state contingency table, so the probability maps express only the likelihood of discovering oil. It is possible, given an adequate historical base, to construct a multistate contingency table that will yield not only the likelihood of discovery, but also the likelihood that discoveries will be of specified magnitude. A series of probability maps can then be drawn, one for each category of field size. The sum of the probabilities on these maps will be equal to the probability maps shown in Figure 5.23 and 5.24. Such an approach provides a way of estimating not only the desirability of a prospective area in terms of anticipated discoveries, but also allows quantitative estimates to be made of the volume of oil that will be found.

CHAPTER 6

Probability Analysis of Multiple Structural Horizons

ANALYSIS OF MULTIPLE GEOLOGIC VARIABLES

The application described in Chapter 5 is concerned with the analysis of only one variable, the structural elevation of a single horizon. Commonly, however, structural information from several horizons is considered simultaneously. The appropriate statistical methods for such studies are those of multivariate analysis, since the basic data involve more than one geologic variable, and interactions between the variables should be considered as well as changes in the variables themselves.

The basic aim of most multivariate procedures is to reduce the complexity of a data set to a simple statistical structure that best "explains" the major portion of the observed variation. One of the best-known procedures is factor analysis, which mathematically transforms a set of observed variables such as structural tops or lithologic characteristics into a smaller set of explanatory "factors." These factors may be identified as hypothetical geologic controls such as "structure," "permeability," and so on. A host of alternative multivariate statistical methods are available whose aims are generally similar but which approach the problem from different mathematical viewpoints. Some useful texts from the vast literature of multivariate statistics include Kendall (1968), Kendall and Stuart (1967), Anderson (1958), Overall and Klett (1972), Morrison (1967), Cooley and Lohnes (1971), and Marriott (1974). Multivariate procedures are discussed in most geomathematics texts, particularly Harbaugh and Merriam (1968), Davis (1973), Agterberg (1974), Blackith and Reyment (1971), and Koch and Link (1970).

DISCRIMINANT FUNCTIONS FOR PROBABILISTIC ASSESSMENT

The most appropriate statistical method for applications in oil exploration should incorporate the historical experience contained in the outcomes of the wells where the subsurface variables are measured. If the well outcomes are linked with multiple geologic variables, a multivariate analysis may be used to select the best combination of variables which differentiate producing areas from dry areas. Hindsight analysis of geological information from these wells may then be used as a tool to predict the potential of new prospects. An effective multivariate method for this purpose is discriminant function analysis, a technique originally developed by Fisher (1936). We illustrate the use of multiple geologic variables and discriminant function analysis with a study by Abry (1973, 1975) which involves the analysis of two structural horizons in the Tatum basin of New Mexico.

Geology of the Tatum Basin

The Tatum basin of southeastern New Mexico is a part of the Permian basin and is located at the northern end of the Delaware basin (Figure 6.1). Abry examined a 37 × 41 mile rectangular area in the northern part

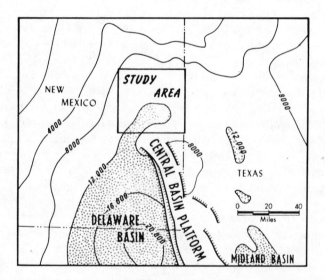

FIGURE 6.1.   Index map of part of Permian basin showing location of study area in Tatum basin of New Mexico. Structure contours showing elevation of top of Precambrian basement are also shown. Structure contours are in feet below sea level. After Abry (1975).

of Lea County, New Mexico, just beyond the north end of the Central basin platform. The Tatum basin is a relatively mature oil province, and the number of wells drilled is adequate for reliable statistical analysis. Intensive exploration began in 1949, providing an historical record from which "hindsight" experiments can be conducted.

In the Tatum basin, rocks of Paleozoic, Triassic, and Tertiary age are present (Figure 6.2). The maximum combined thickness of Paleozoic strata is more than 15,000 feet in the deepest part of the basin. Abry's study was confined to oil production from Silurian and Devonian units, although he utilized structural information from the Permian Abo Formation. The 21 oil fields producing from the Siluro-Devonian in the study area are listed in Table 6.1; their sizes vary from 0.5 to 125 million barrels of total recoverable oil and gas equivalent to oil. Figure 6.3 shows the outlines of the fields as of 1970.

The Siluro-Devonian section consists of locally cherty dolomitic limestones that were deposited in clear, shallow seas which covered a broad southward sloping submerged shelf (Galley, 1958). Although Devonian

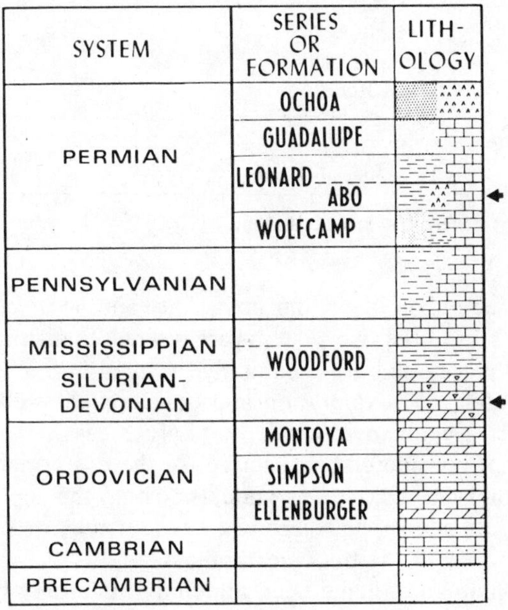

SYSTEM	SERIES OR FORMATION	LITH-OLOGY
PERMIAN	OCHOA	
	GUADALUPE	
	LEONARD / ABO	
	WOLFCAMP	
PENNSYLVANIAN		
MISSISSIPPIAN	WOODFORD	
SILURIAN-DEVONIAN		
ORDOVICIAN	MONTOYA	
	SIMPSON	
	ELLENBURGER	
CAMBRIAN		
PRECAMBRIAN		

FIGURE 6.2. Generalized stratigraphic and lithologic column for Permian basin (after Galley, 1958). Siluro-Devonian (lower arrow) is producing interval in Tatum basin, although Abry used structure of Permian Abo Formation (upper arrow) to forecast structure in Siluro-Devonian.

TABLE 6.1. Oil fields that produce from Siluro-Devonian in Abry's study area in Tatum basin.

Discovery Year	Field Name	Total Recoverable Reserves (Millions of Barrels)
1949	Bagley	25.0
1950	Gladiola	65.0
1950	Denton	125.0
1951	Echol	5.0
1951	E. Caprock	25.0
1952	*Bronco	25.0
1952	N. Echol	1.2
1952	Moore	20.0
1952	Mescalero	4.0
1953	*Anderson Ranch	8.0
1954	Caudill	5.0
1954	S. Cross Road	4.0
1955	S. Denton	3.5
1955	Dean	2.7
1956	Gross	0.5
1956	King	6.0
1956	Four Lake	1.7
1957	E. Echol	0.5
1959	High Tower	1.0
1960	S. Gladiola	4.0
1961	*Medicine Rock	2.0

*Fields at edge of area not included in study.

rocks contain abundant chert and novaculite and lie disconformably on the Silurian, rocks of the two systems are difficult to distinguish lithologically in many places and are commonly grouped into a single unit.

Within the area, 653 exploration and development wells partly penetrate the Siluro-Devonian interval, but only 3 reach its base. Deeper horizons are not considered prospective, so there is no incentive to drill through the interval. As a rule, wildcats testing the top of the Siluro-Devonian are abandoned unless oil is encountered, in which case the wells are deepened to the oil-water contact.

Siluro-Devonian strata have been affected structurally by movements during several geologic periods, notably in early Pennsylvanian and again in early Permian time. The resulting structural style is one of fault blocks separated by nearly vertical fault planes. Most oil and gas is trapped in fold and fault structures and along truncated edges of beds that form

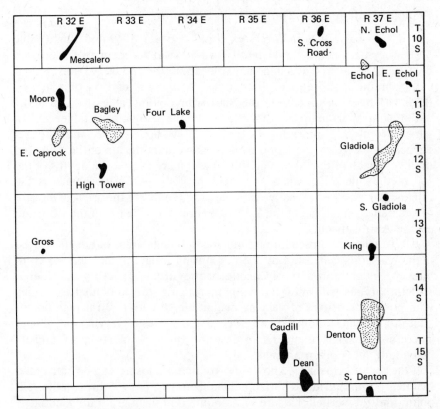

FIGURE 6.3.   Oil fields in Tatum basin study area in 1970. Fields discovered before end of 1951 are solid black, whereas fields discovered afterwards are stippled. Adapted from Abry (1973).

buried hills below the unconformity at the top of the Siluro-Devonian. Exploration strategy consists of seeking anticlines by subsurface mapping based on well-log correlations, and by seismic exploration.

The Siluro-Devonian strata probably were exposed in late Silurian time and also prior to deposition of the Woodford Shale. During these times of exposure they underwent solution, leaching, and dolomitization. In some areas, particularly on anticlines, part of the Siluro-Devonian interval is missing. The entire Siluro-Devonian in Abry's area acts as a single reservoir, although to the south permeable zones in the Siluro-Devonian are separated by several impermeable horizons. Reservoir porosity consists of vugs and intercrystalline pores and ranges up to 15 percent. Permeability is poor, commonly less than 100 millidarcies.

Statistical Analysis of Tatum Basin Structure

The Tatum basin was intensely explored between 1949 and 1962 (Table 6.2). Abry designed a "hindsight" predictive experiment that used well data available at an early stage of exploration (the end of 1951) to statistically "predict" subsequent discoveries. At the end of 1951 only five oil fields had been discovered in the Siluro-Devonian (Tables 6.2 and 6.3).

For statistical analysis Abry selected nine geologic variables, all of which are closely interrelated (Table 6.4). Variables 1, 2, and 3 reflect vertical closure on the top of the Siluro-Devonian, the top of the Abo, and on the isopach of the Abo. Variables 4 and 5 are the subsea depths of the top of the Siluro-Devonian and the top of the Abo. Variable 6 is the thickness of the Abo. Finally, variables 7, 8, and 9 are the areas of closure of anomalies on the top of the Siluro-Devonian, on the top of the Abo, and on the Abo isopach.

Abry designed a procedure to find and delineate areas of closure on the numerical grids generated by the contouring program, and then to compute the height and area of closure of the structures. Abry considered structural types as forming a continuum ranging from anticlinal anomalies with positive closures of varying magnitude (Figure 6.4) through homoclines to synclinal anomalies with negative closures of varying magnitude. Therefore, a single number can describe the type of structure and its magnitude for a given anomaly.

If the area of closure is known the volume of closure can be calculated by integration. A gridded horizontal plane may be passed through the spill point, and the vertical closure, h, computed at each grid intersection. If the area of a grid element is a and the structure covers n grid points (Figure 6.5), the volume of closure, V, is approximated by the sum of the heights of closure multiplied by the area per grid cell:

$$V = a \sum_{i=1}^{n} h_i.$$

A search was made on the structural grid to find the top of an anomaly, and then a search was made around the structure top at an increasing distance until the spill point of the structure was found. Abry distinguished two situations concerning the relationships between spill point and reservoir rock thickness:

1. The plane of the lowest possible oil-water contact does not intersect the base of the reservoir rock (Figure 6.6a). Then the vertical closure at a particular location within the anomaly is equal to the difference in eleva-

TABLE 6.2. Wells drilled from 1936 to 1962 in Abry's Tatum basin study area, and names of oil fields producing from Siluro-Devonian. The number of wells drilled in the area since 1962 is relatively small.

Year	Number of Wells Penetrating Top of Siluro-Devonian		Number of Wells Penetrating Abo (Permian)		Fields Discovered in Siluro-Devonian
	Annual	Cumulative	Annual	Cumulative	
Wells providing information for prediction					
1936	1	1	1	1	
1940	1	2	1	2	
1944	0	2	1	3	
1948	2	4	2	5	
1949	12	16	17	22	Bagley
1950	24	40	32	54	Gladiola, Denton
1951	71	111	128	182	Echol, E. Caprock
Wells providing information to test predictions					
1952	71	182	167	349	*Bronco, N. Echol, Moore, Mescalero
1953	77	259	181	530	*Anderson Ranch
1954	42	301	105	635	Caudill, S. Cross Road
1955	48	349	96	731	S. Denton, Dean
1956	71	420	115	846	Gross, King, Four Lake
1957	90	510	121	967	E. Echol
1958	24	534	59	1026	
1959	16	550	58	1084	High Tower
1960	13	563	31	1115	S. Gladiola
1961	19	582	40	1155	*Medicine Rock
1962	22	604	62	1217	

*Fields at edge of area not included in study.

TABLE 6.3. Siluro-Devonian oil fields in Tatum basin study area at end of 1951, classified by size.

Field	Class	Class Size (Millions of Barrels)	Actual Field Size (Millions of Barrels)
Dry and noneconomic wells	A	<1 or dry	<1
Echol	B	1 to 26	5
E. Caprock			25
Bagley			25
Gladiola	C	>26	65
Denton			125

TABLE 6.4. Geologic variables chosen by Abry for statistical oil-occurrence model.

Variable	
1	Closure of structural anomalies of Siluro-Devonian top
2	Closure of structural anomalies of Abo top
3	Closure of Abo isopach
4	Depth of Siluro-Devonian top (subsea)
5	Depth of Abo top (subsea)
6	Isopach value of Abo
7	Area of closure of structural anomalies on Siluro-Devonian top
8	Area of closure of structural anomalies on Abo top
9	Area of closure of structural anomalies on Abo isopach

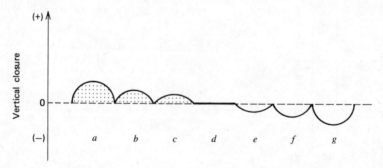

FIGURE 6.4. Relative scale of fold structures. Anticlines (a, b, c) have positive closure; homocline (d) has zero closure; and synclines (e, f, g) have negative closure. After Abry (1975).

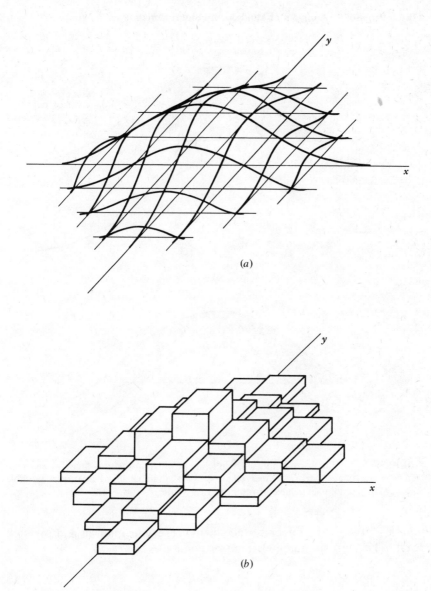

FIGURE 6.5. (*a*) Closed anticlinal structure represented as contour grid matrix, and (*b*) its representation as aggregation of columns of different height. After Abry (1973).

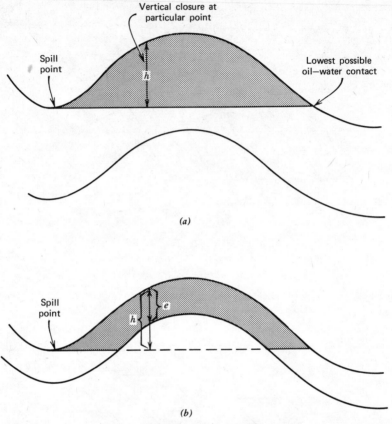

FIGURE 6.6. (*a*) Cross-section through anticline in which lowest possible oil-water contact does not intersect base of reservoir rock. (*b*) Anticline in which lowest possible oil-water contact intersects base of reservoir rock. After Abry (1973).

tion, *h,* between the upper surface of the reservoir interval and the spill point. The corresponding element of volume is:

$$\Delta V = ah$$

2. If the spill point elevation intersects the base of the reservoir rock (Figure 6.6*b*), the vertical distance between the top and bottom of the reservoir, *e*, is less than the vertical closure, *h*. The element of volume, ΔV, is

$$\Delta V = ae$$

In a similar way the heights and areas of "negative closure" of basins and synclines can be found. A search begins at a grid intersection surrounded by grid points of higher elevation. Heights and areas of closure, however, were treated by Abry as negative values to indicate that they describe synclines. Homoclines were assigned a height and area of closure of zero. In Abry's study, height of closure and area of closure were distinct instead of being combined into volume of closure, to differentiate between anomalies having similar volumes but different shapes (Figure 6.7). Examples of maps showing areas and maximum heights of closure are shown in Figures 6.8 and 6.9.

Discriminant Function Analysis

For each wildcat well, Abry attempted to predict the outcome in terms of the three production classes of Table 6.3, on the basis of the nine geologic variables given in Table 6.4. Assignments to the three classes were based on the geology of the well sites as perceived prior to drilling. Classification into predefined categories was done by discriminant analysis, a procedure introduced in 1936 by R. A. Fisher, and subsequently used with success in many geological studies where it was first introduced by Emery (1954); for other examples, see Middleton (1962), Chayes (1964), and Krumbein and Graybill (1965). The procedure is used both as a means to characterize predefined populations and as a tool to classify individuals of unknown affinities.

The simplest and most common application of discriminant analysis is to find the linear combination of variables that best distinguishes two groups. Two initial sets of samples must be assigned to the two classes, each of which corresponds to distinctive populations. The initial assignment must be made on the basis of prior knowledge.

If k variables are measured on each observation, the two sets of samples may be plotted as two clusters of points in k-dimensional space, where each axis corresponds to a variable. Within this k-dimensional space, it is possible to locate an axis on which the distance between the

FIGURE 6.7. Anticlinal structures containing similar volumes of oil but with different combinations of heights and areas of closure. Adapted from Abry (1973).

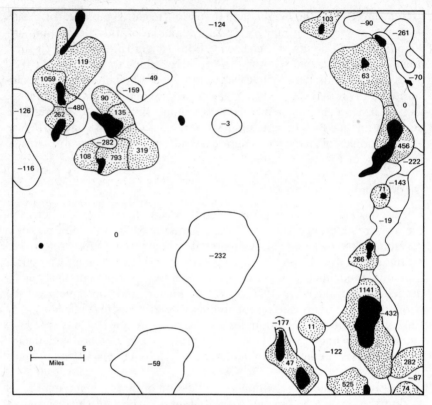

FIGURE 6.8.   Map of top of Siluro-Devonian in Abry's study area displaying heights of closure. Area has been analyzed on grid with 39 columns and 35 rows of cells, each one mile square. Maximum height of closure of structure, expressed in feet, has been assigned to all grid points within each anomaly. Values are positive for anticlinal structures, negative for synclinal structures, and zero for homoclines. Redrawn from computer printout. Oil fields are solid black; positive anomalies are stippled. Adapted from Abry (1973).

centers of the two sample clusters is maximized and simultaneously the dispersion within each cluster is minimized. This axis defines the *linear discriminant function,* which is an equation derived from the multivariate means, variances, and covariances of the two clusters (for mathematical background, see Davis, 1973, pp. 442–456). Substitution of the midpoint between the two samples' means in the discriminant function yields a *discriminant index* which defines a partition point along the discriminant function line. *Discriminant scores* may be calculated for individual observations by projecting their multivariate coordinates onto the discriminant function. The scores are used to classify individuals into one or the other

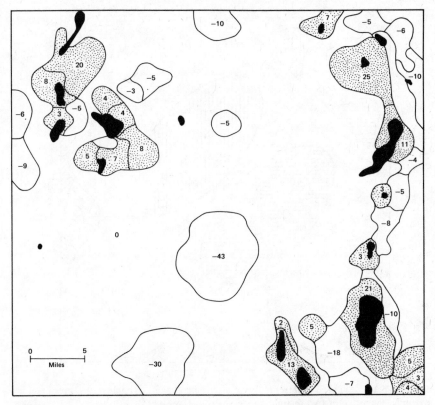

FIGURE 6.9. Map corresponding to Figure 6.8 except that areas of closed anomalies are shown instead of maximum height. Numbers indicate area of closure in square miles in each anomaly. Negative values denote "negative closure" in synclines and closed basins. Positive anomalies are stippled. Adapted from Abry (1973).

of the two clusters, depending on which side of the discriminant index they lie. The difference between a discriminant score and the discriminant index reflects the confidence with which an individual can be assigned to a cluster.

The initial assumption or hypothesis that the two samples really belong to distinct populations may be judged by an F-test utilizing *Mahalanobis' generalized distance,* which is a measure of the distinctness between the clusters in k-dimensional space. The statistical test has an associated level of significance, which is the probability of concluding that two populations exist when actually there is only one. Certain statistical assumptions must be satisfied by the data if the test is to be valid.

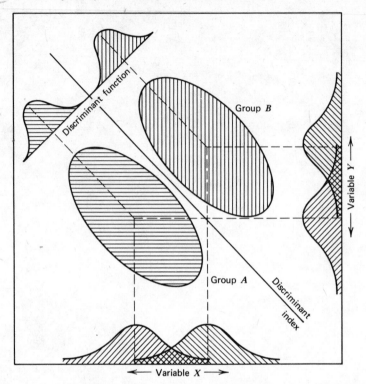

FIGURE 6.10.   Pictorial representation of discriminant analysis applied to simple bivariate example consisting of two distinct clusters. From Davis (1973).

The two hypothetical populations shown in Figure 6.10 cannot be successfully distinguished by a single variable. When plotted in two dimensions, however, a distinct separation is apparent and a discriminant function can be computed that will distinguish the two groups. This example is idealized in that a perfect separation is possible; in realistic situations the clusters may overlap across the discriminant index, and sample observations must be assigned to one or the other of the clusters as matters of probability.

Discriminant function analysis is readily extended to segregation of three or more groups. Figure 6.11 represents three clusters whose members are defined by two variables. Notice that clusters *A, B,* and *C* are not distinct on the basis of either variable alone, because the clusters overlap strongly on either axis.

A transformation of the original variables into *canonical variables* can be made, so that the first new canonical axis is inclined in the direction of greatest variability between the group means. The second axis is perpen-

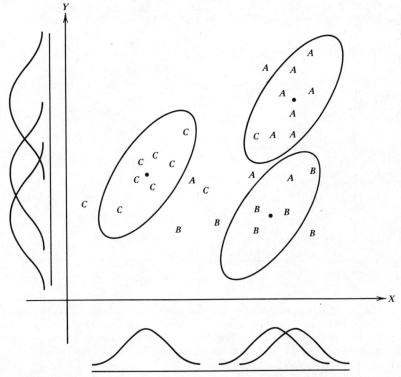

FIGURE 6.11. Plot of three different clusters A, B, and C, as function of two variables X and Y. Note that probability density functions of three groups overlap on both axes, and therefore variables X and Y in their original form are not sufficient to distinguish three clusters.

dicular to the first, and inclined in the direction of next greatest variability (Figure 6.12). Detailed explanations of the computation of these canonical variables are given in texts such as Blackith and Reyment (1971) and Marriott (1974). The multigroup discriminant functions are planes that bisect the segments joining the multivariate means of the groups, and the portions of the space thus cut off are the *domains* of the different groups. Additional samples can be assigned to a particular group on the basis of their discriminant scores depending on that portion of the sample space into which they fall.

Assignment of Control Samples to Clusters

The likelihood that a sample of unknown origin belongs to a particular cluster decreases with the distance between the sample and the mean of

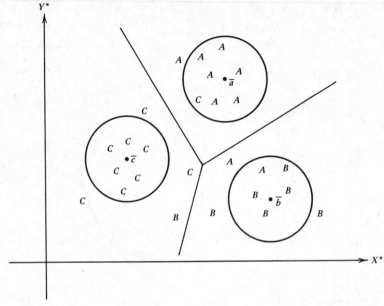

FIGURE 6.12.   Plot of same population in Figure 6.11, except that original variables X and Y have been transformed into new (canonical) variables, X^* and Y^*, which have variance of one and are uncorrelated. Discriminant functions are bisectors of segments joining bivariate means of clusters $\bar{a}$, $\bar{b}$, and $\bar{c}$. Three clusters are now more distinct on basis of discriminant functions, but minor misclassifications remain.

the cluster considered (Figure 6.12). For example, any sample falling on the discriminant function line between clusters A and B and far from cluster C has a likelihood of belonging to either cluster A or B. A sample which falls on the line joining the bivariate means $\bar{a}$, $\bar{b}$ has a greater likelihood of belonging to cluster A if its distance from $\bar{b}$ is greater. Beyond point $\bar{a}$, the distance between the sample and $\bar{b}$ increases, but so does the distance between the sample and $\bar{a}$. Then comes a point after which the likelihood that the sample belongs to either cluster decreases. When the distances to all cluster means are large, forcing an assignment of a sample to one of them would be unsatisfactory. Such a sample is not assigned if the distances to all cluster means are greater than a threshold distance determined by examining the overall distances between the control samples.

Abry used discriminant function analysis to assign proposed wildcat locations to one of the three production classes of Table 6.3. Estimates of the nine geologic variables obtained by interpolation from computer-contoured maps were entered into the discriminant classification model and the likelihood of correct classification into one of the three clusters was computed. The assumption was made that the relationships between oil occurrence and geologic factors do not vary abruptly over the study area, and that changes in perceived geology have been gradual as the basin has become more thoroughly explored. The initial assignments of the control or "training" samples (well locations) to the three classes of field sizes (groups A, B, and C) are listed in Table 6.5. Group D includes prospective well sites that have not yet been drilled and are to be classified into groups A, B, and C. In other words, the control wells serve as training groups to calculate discriminant functions, which in turn permit undrilled locations to be assigned to one of the three outcome categories. Initially the control samples were classified using all nine geologic variables. However, analysis showed that only four geologic variables were sufficient to categorize the undrilled locations.

To test the efficacy of the discriminant function, values of control samples at locations of previously drilled wells were entered into the discriminant equation and their discriminant scores calculated. Results are shown in Table 6.6; all but one well (number 7 of group A) were correctly classified. Thus, at least among the control samples, locations can be successfully placed in one of three outcome categories solely on the basis of the geologic variables.

Identification of Important Geologic Variables

Abry used a form of discriminant analysis called a *stepwise discriminant function* in which the calculations were not performed on all variables simultaneously but instead on individual variables taken consecutively in order of decreasing power of discrimination. This permitted the relative importance of each geologic variable to be determined. Following initial

TABLE 6.5. Sample assignments to groups A, B, C, and D. Control samples are wells drilled before end of 1951 and placed in size class A, B, and C. Unclassified samples (D) are undrilled grid locations, each of which was subsequently assigned to either class A, B, or C by use of discriminant analysis.

Group Label	Class (Millions of Barrels)	Number of Samples in Group	Type of Sample
A	<1 or dry	14	Control
B	1 to 26	10	Control
C	>26	13	Control
D	Unknown	1328	Unclassified

TABLE 6.6 Reassignment of control samples (wells drilled before end of 1951) on basis of discriminant analysis of nine geologic variables. After Abry (1973).

Class to Which Control Well Was Initially Assigned	Class to Which Each Control Well Was Reassigned	Square of Distance (M) and Likelihood Value (L) Calculated for Each Control Well					
		Group A		Group B		Group C	
		M	L	M	L	M	L
Group A wells							
1	A	2.4	0.9	7.9	0.0	82.0	0.0
2	A	11.8	1.0	27.9	0.0	87.8	• 0.0
3	A	3.7	0.9	9.4	0.0	71.1	0.0
4	A	5.6	0.9	12.5	0.0	74.9	0.0
5	A	1.8	0.9	12.3	0.0	88.4	0.0
6	A	11.8	0.8	14.7	0.1	72.2	0.0
7	B	9.0	0.0	3.9	0.9	84.7	0.0
8	A	3.3	1.0	21.6	0.0	77.0	0.0
9	A	17.4	1.0	43.9	0.0	113.8	0.0
10	A	26.6	0.9	39.5	0.0	113.0	0.0
11	A	5.6	1.0	22.0	0.0	84.3	0.0
12	A	8.1	1.0	27.3	0.0	87.7	0.0
13	A	10.5	1.0	40.6	0.0	82.3	0.0
14	A	14.1	0.9	28.7	0.0	74.0	0.0
Group B wells							
1	B	21.9	0.0	5.7	1.0	110.7	0.0
2	B	17.6	0.0	5.1	0.9	81.1	0.0
3	B	18.8	0.0	4.9	0.9	96.7	0.0
4	B	17.6	0.0	4.9	0.9	84.0	0.0
5	B	21.7	0.0	5.9	1.0	117.9	0.0
6	B	25.0	0.0	8.4	1.0	120.7	0.0
7	B	21.1	0.0	7.8	0.9	91.7	0.0
8	B	13.4	0.0	2.2	0.9	81.6	0.0
9	B	13.7	0.0	6.7	0.9	56.0	0.0
10	B	20.8	0.0	11.0	0.9	82.5	0.0
Group C wells							
1	C	83.6	0.0	104.6	0.0	10.1	1.0
2	C	60.9	0.0	72.0	0.0	4.0	1.0
3	C	85.6	0.0	100.2	0.0	6.0	1.0
4	C	58.6	0.0	66.6	0.0	4.3	1.0
5	C	57.9	0.0	65.1	0.0	4.4	1.0
6	C	59.7	0.0	68.7	0.0	4.1	1.0
7	C	91.4	0.0	103.1	0.0	3.0	1.0
8	C	84.5	0.0	106.3	0.0	15.1	1.0
9	C	80.2	0.0	86.1	0.0	7.5	1.0
10	C	63.2	0.0	71.1	0.0	8.5	1.0
11	C	144.7	0.0	153.8	0.0	14.1	1.0
12	C	104.3	0.0	99.9	0.0	10.5	1.0
13	C	120.9	0.0	130.8	0.0	18.3	1.0

calculations, some variables proved to be of marginal usefulness because they did not improve the efficacy of the assignment of training wells to the proper groups. Table 6.7 gives the ranking of the geologic variables in order of usefulness; the most effective is vertical closure of the Siluro-Devonian anomalies and the depth of the top of the Siluro-Devonian is second in importance. Each F-value given in Table 6.7 is a measure of the contribution of the particular geologic variable to the discrimination process. In Abry's example, the first four variables are almost as effective as all nine geologic variables combined.

Classification of Undrilled Grid Locations

Each of Abry's unclassified prospect locations in group D was assigned to either group A, B, or C after computing the Mahalanobis' distances to the groups and likelihood values with respect to each group. Table 6.8 shows the number of control samples assigned to groups A, B, and C, and the number of unclassified locations assigned to the same groups. Of the prospective locations, 957 were subsequently assigned to group A, 247 to group B, and 124 to group C. However, not all of these assignments are meaningful as some samples are situated on the extreme edges of the study area. These locations are characterized by large Mahalanobis' distances to all groups, and assignment to a specific class is dubious. The threshold arbitrarily chosen for meaningful Mahalanobis' distances was

TABLE 6.7. Ranking of geologic variables in stepwise discriminant analysis according to their contribution to discrimination between three outcome classes defined in Table 6.3. From Abry (1973).

Variable	F-value
Vertical closure of top of Siluro-Devonian	61.0158**
Siluro-Devonian subsea depth	20.1555**
Vertical closure on Abo isopach	16.3285**
Abo isopach value	7.7582**
Area of closure on Abo isopach	4.9897*
Area of closure on Siluro-Devonian	3.8013
Vertical closure on Abo top	1.8973
Abo subsea depth	0.5881
Area of closure on Abo top	0.3946

*Significant at 95% level, with 1 and 27 degrees of freedom.
**Significant at 99% level, with 1 and 27 degrees of freedom.

TABLE 6.8. Number of control samples (wells) from groups A, B, C and grid locations of unclassified samples (group D) subsequently assigned to groups A, B, and C on the basis of the geologic factors.

Original Group	Group to Which Samples (Wells or Grid Locations) Were Subsequently Assigned		
	A	B	C
A 14	13	1	0
B 10	0	10	0
C 13	0	0	13
D 1328	957	247	124

100, and locations above this value are shown with zero likelihood values on the maps which follow.

To facilitate interpretation, results of the classification of the undrilled grid locations are shown in Figure 6.13. At each location, the class assignment with the highest likelihood is listed, provided the likelihood is 50 percent or greater. The likelihood values for assignments of prospective localities to classes A, B, and C are contoured in Figures 6.14 to 6.16. These maps readily demonstrate the potential of the method.

TABLE 6.9. Results of oil-field class assignments for fields discovered after 1951 based on geologic data available at end of 1951. From Abry (1973).

Field	Field Size (Millions of Barrels)	Actual Group	Assigned by Discriminant Analysis to Group
Caudill	5.0	B	B
Dean	2.7	B	B
S. Denton	3.5	B	B
King	6.0	B	B
Moore	20.0	B	C
High Tower	1.0	B	C
S. Gladiola	4.0	B	B
Mescalero	4.0	B	B
Four Lake	1.7	B	A
N. Echol	5.0	B	A
E. Echol	0.5	A	A

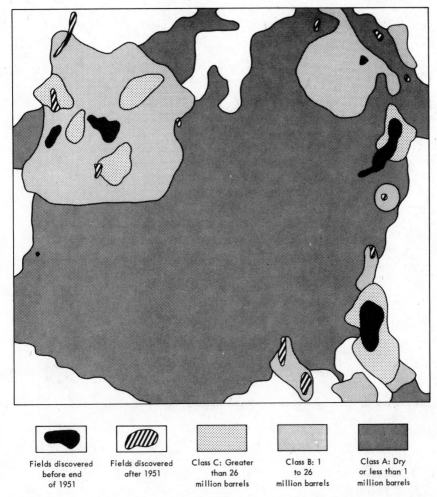

FIGURE 6.13. Map showing outcome classes established by discriminant function analysis of nine geologic variables. If feasible, an outcome class was established for each of 1365 cells forming grid of 39 columns and 35 rows. Map patterns outline geographic groups of cells in each particular class. Areas containing cells which could not be readily assigned to given class are blank. Adapted from Abry (1973).

Fields discovered before end of 1951

Fields discovered after 1951

Class C: Greater than 26 million barrels

Class B: 1 to 26 million barrels

Class A: Dry or less than 1 million barrels

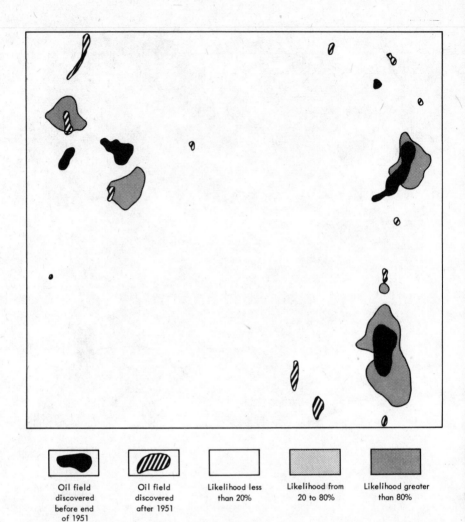

FIGURE 6.14. Map showing likelihood of correct assignment to class C (fields larger than 26 million barrels) of grid locations that had not been drilled as of December 31, 1951. Fields discovered before end of 1951 are solid black, whereas fields discovered after 1951 are stippled. Data from Abry (1973).

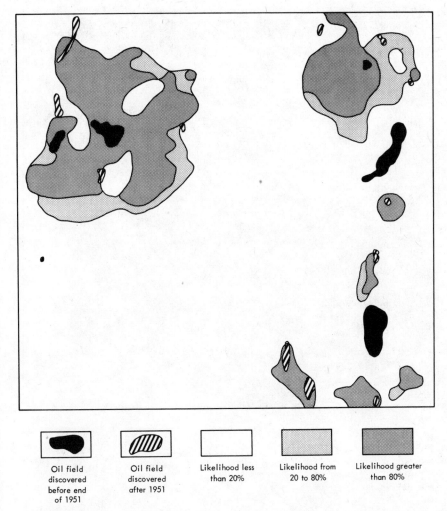

FIGURE 6.15. Map showing likelihood of correct assignment to class *B* (fields between 1 and 26 million barrels) of grid locations that had not been drilled as of December 31, 1951. Fields discovered before end of 1951 are solid black, whereas fields discovered after 1951 are stippled. Data from Abry (1973).

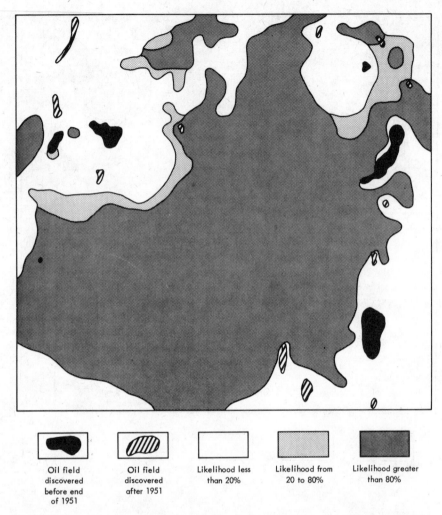

FIGURE 6.16. Map showing likelihood of correct assignment to class *A* (dry or oil fields smaller than 1 million barrels). For each grid location, the sum of assignment likelihoods for all three classes (Figures 6.14 to 6.16) is 100 percent. Data from Abry (1973).

Oil field discovered before end of 1951

Oil field discovered after 1951

Likelihood less than 20%

Likelihood from 20 to 80%

Likelihood greater than 80%

The results of classification of oil fields discovered after 1951 on the basis of pre-1952 information are summarized in Table 6.9. Of the 13 fields that lie within the study area, it was possible to classify 11 of them. The remaining two (Gross and S. Cross Road) could not be classified because no wells had been drilled nearby before 1952. Of the 11 fields classified, 7 are classified correctly in size categories A and B and the remaining 4 were misclassified.

In Abry's study, information obtained before 1952 from five oil fields and a number of dry wells permitted formulation of an effective oil-occurrence model, which correctly identified the locations of seven oil fields that were discovered after 1951. In addition, there are locations not within the boundaries of known fields when Abry completed his study in 1973, which were classified as being within Group B (1 to 26 million barrels) or within group C (greater than 26 million barrels). These localities appear to warrant additional exploration.

CHAPTER 7

Probabilistic Exploration for Stratigraphic Traps

ANALYSIS OF COMBINED STRUCTURAL AND STRATIGRAPHIC FEATURES

Up to this point our analysis has been restricted to relationships between oil accumulation and structural measures of the producing formation. Knowledge of these relationships is usually sufficient in exploring for purely structural traps when the location of oil fields is influenced primarily by buoyant entrapment of hydrocarbons in local positive structures. Stratigraphic traps, however, represent an increase in complexity since several geological variables must be studied simultaneously in order to pinpoint trap locations effectively. In statistical terms, definition of stratigraphic trap locations is a multivariate problem because the internal characteristics of the producing formation must be considered as well as structural configuration.

The definition of stratigraphic traps is subtle in the sense that several key steps are involved, each of which requires judgment and geological expertise. First, the geological variables that influence stratigraphic oil entrapment must be selected and must be simultaneously considered over the exploratory area. Then, each of these variables must be weighted in terms of its contribution to entrapment to yield a composite evaluation of any location of interest.

Exploration for stratigraphic traps involves blending practical experience with inferences drawn from geological theory. On the basis of observations of oil field/subsurface geological relationships ("experience"), the petroleum geologist selects diagnostic formation properties and maps these over an area. By comparing various maps he may discern the multiple relationships between known fields and a combination of geologic variables and then look for similar relationships elsewhere which may indicate prospects. Some of the geologic variables or combinations of variables may seem to be critical; others may appear to be of doubtful importance or even irrelevant. As in the previous studies, the selection of prospects represents an extrapolation into the unknown. Predictions must be tempered by the degree of available well control and the apparent scale of variation in the diagnostic variables.

The conventional strategy described may be paralleled by numerical methods developed in multivariate statistics. The advantages of a mathematical approach are that relationships between field locations and subsurface geology may be more objectively analyzed, and that the analysis may be based on a large number of variables used as an integrated whole, with an understanding of each variable's unique contribution to the statistical prediction of trapping situations. These features are an advance over normal human capability in which truly objective judgments are difficult and assessing more than a few factors simultaneously is virtually impossible. By contrast, however, a human has a greater flair for pattern recognition, coupled with a greater flexibility in analysis. Since the strengths and weaknesses of the two approaches largely complement one another, the interactive use of statistical modeling and human selectivity appears highly desirable.

EVALUATION OF STRUCTURAL-STRATIGRAPHIC PROSPECTS IN SOUTH-CENTRAL KANSAS

In south-central Kansas, major gas fields are contained in weathered chert developments of the Mississippian Osagian Stage and are deployed in a loose peripheral arc south of the Central Kansas Uplift (Figure 7.1). Entrapment of gas appears to be controlled by lateral permeability changes, structural features, and pinchouts of the truncated edge of the Osage, which is overlapped by the basal Pennsylvanian unconformity.

The stratigraphic unit studied in this case history is the *B* division of the Osage, an informal lower subdivision that is separated from the Mississippian *A* by shale (Figure 7.2). A 13 × 13 mile study area was selected in southwest Stafford County and lies to the north of the major gas fields mentioned above. The area straddles the southwestern margin of the Central Kansas Uplift and the northern extension of the Pratt Anticline, two major structural elements whose early Pennsylvanian

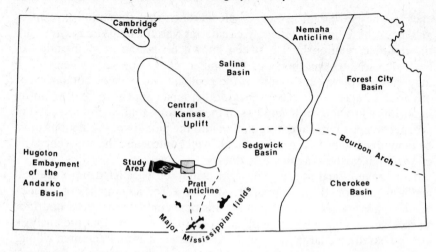

FIGURE 7.1.   Location of study area and major tectonic elements of Kansas (adapted from Merriam, 1963).

movements strongly affected both structural and compositional features of the Osage. Within the study area, the Osage subcrops to the north and is truncated and overlapped by the basal Pennsylvanian conglomerate which rests successively on rocks of the Kinderhookian Stage and the Viola Limestone (Figure 7.3).

Extensive upward movement of the Central Kansas Uplift in late Mississippian and early Pennsylvanian time led to erosion of Mississippian and older Paleozoic rocks from the higher part of the structure (Merriam, 1963). The uplift also initiated local structures that are probably related to differential movement along old zones of weakness. The grain of the local structure is marked by two trends, one northeast-southwest parallel to the Nemaha Anticline, and the other northwest-southeast parallel to the axis of the Central Kansas Uplift. The interplay between tectonic structures and features possibly caused by subaerial weathering has created a complex Osagian surface, reflecting both structural and ancient topographic elements (Clair, 1948; Beebe, 1959). Weathering or modification by percolating water has produced a thick residuum of weathered chert that flanks the southern margin of the Central Kansas Uplift as a broad arcuate rim.

Figure 7.2 shows a typical stratigraphic section through the Osage in the study area, based on core descriptions from the Western Petroleum #1 Hart well which produces from the Mississippian *B* and lies within the Haynes field. Selected fragments of core are shown in Figure 7.4. The

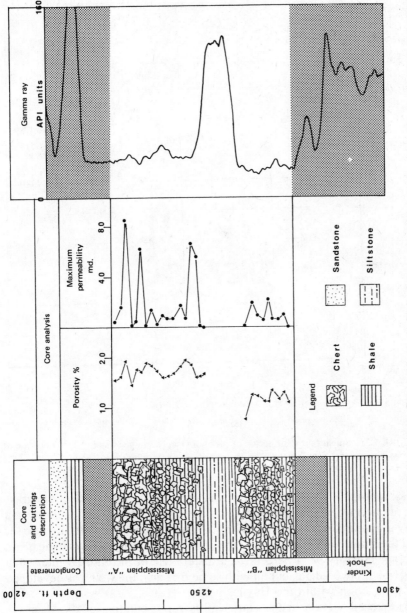

FIGURE 7.2. Characteristics of Mississippian System penetrated by Western Petroleum Hart #1 well (C-SE-NW 22-25S-15W).

195

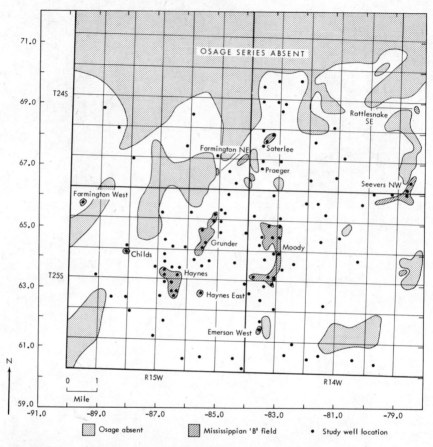

FIGURE 7.3. Subsurface distribution of rocks of Osagian Stage in study area. Location of Mississippian *B* fields and wells used for statistical analysis are also shown.

Mississippian *B* division contains coarse to fine, irregular, white to gray, pitted chert fragments that tend to dovetail together and are enclosed in a limy gray-green shale matrix. Within the unit, chert fragments are coarser and the shale content is lower in the upper part. Core analyses indicate an average porosity of 11 percent. A drillstem test of the entire *B* interval in this well recovered gas-cut oil with some free gas; the well was developed for production within the Haynes field.

The cored section in the Hart well is representative of the Osage in Stafford County. Although the *A* and *B* subdivisions can be traced across the area, correlation becomes more difficult near the regional edge of the Osage and near local inliers, where the Osage has been removed by

SUBSURFACE DEPTH (FEET)

4256   4263   4283

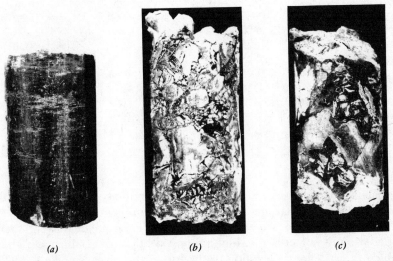

(a)   *(b)*   *(c)*

FIGURE 7.4. Core samples from Western Petroleum Hart #1 well. (*a*) Shale overlying Mississippian *B*. (*b*) and (*c*) Top and bottom, respectively, of Mississippian *B*.

erosion. Well logs, cores, and samples show that the proportion of chert and shale in the Mississippian *B* unit varies systematically across the area.

A number of oil and gas fields occur in the Mississippian *B* in the study area (Figure 7.3 and Table 7.1), although they are smaller than the major gas fields to the south. Exploration experience in Stafford County suggests that Mississippian *B* production is most common in areas in which the clay content of the unit is low. The best trapping sites involve favorable local structures, although structural closure is not necessarily a key factor.

Selection and Measurement of Geological Variables

It is likely that the location of fields in the Mississippian *B* of southwestern Stafford County reflects both stratigraphic and structural controls. Lateral gradations in the chert/shale ratio cause significant variations in permeability, with a potential for stratigraphic traps within the unit, in addition to those located at the truncated margins of the Osage. Local closure also may have created purely structural traps within more

TABLE 7.1.   Mississippian *B* oil and gas fields in study area.

Field	Discovery Well	Discovery Date	Production
Childs	17-25S-15W	1959	Gas
Emerson West	30-25S-14W	1963	Gas
Farmington NE	35-24S-15W	1964	Gas/Oil
Farmington West	6-25S-15W	1963	Gas
Grunder	11-25S-15W	1958	Oil
Haynes	22-25S-15W	1959	Oil
Haynes East	23-25S-15W	1960	Oil
Moody	7-24S-14W	1956	Oil
Praeger	1-24S-14W	1959	Oil
Rattlesnake SE	13-24S-14W	1959	Oil
Saterlee	25-24S-15W	1958	Oil
Seevers NW	1-25S-14W	1958	Gas/Oil

permeable areas. Elsewhere, structural variations contribute to the geometry of traps that are otherwise principally stratigraphic.

Subsurface variables descriptive of the Mississippian *B* were therefore chosen to reflect both structural and stratigraphic influences. These include the structural elevation of the top of the Mississippian *B*, thickness of the interval, and an estimate of its shale content. These three geologic variables were measured from the logs of 124 wells that penetrate the Mississippian *B* in the study area. Since the *B* interval consists of two main lithologies, chert and shale, an estimate of its shale content expresses its gross lithologic character.

Visual observations of drill cuttings and core samples would provide a direct estimate of shale content, but such estimates are more susceptible to individual interpretation by different geologists than a measure from a downhole logging tool. Estimates made from cuttings are also ambiguous because of infiltration by shale cavings derived from shallower horizons. Consequently, the most satisfactory method of estimating shale content is by analysis of geophysical well logs.

The shale content in the Mississippian *B* of each well was estimated by the Century Geophysical Corporation using digitized gamma ray log traces. A mean gamma ray value was computed from a series of discrete counts recorded in API units on the log of each well. All wells drilled prior to 1964 for which gamma ray logs exist were used, enabling the shale content to be consistently compared from well to well.

The gamma ray log is primarily a measure of the concentration of the potassium-40 isotope which occurs in significantly greater quantities in shale than in other common sedimentary rocks. In logging typical sedimentary sequences, the gamma ray response provides an approximate measure of shale content. Since gamma ray logs are calibrated in different ways by logging service companies and are also influenced by borehole conditions, it is necessary to standardize each well log internally by the procedure indicated in Figure 7.5. A value corresponding to an end member of "pure" shale was obtained from an average of the 20 highest gamma ray readings in the overlying Pennsylvanian rocks within an interval extending from 50 feet below the top of the Heebner Shale (Virgilian) to the top of the Lansing Group (Missourian). Similarly, a gamma ray reading for a "clean" formation was obtained by averaging the 20 lowest values recorded in the interval between the top of the Lansing and the base of the Kansas City Group (Missourian). Use of these boundary values assumes that the gamma ray deflection extremes of these intervals are essentially uniform across the area, and that they do not deviate anomalously from either the pure shale or the clean formation they are used to represent. By setting one of the boundary value readings as "zero" (clean formation) and the other as "one" (shale), the average gamma ray deflection in the Mississippian B interval can be used to estimate the proportion of shale within the unit. This procedure is commonly used by log analysts to obtain a "shale correction factor" for the lithologic interpretation of log combinations.

Shale content appears to strongly influence permeability. The fundamental relationship between porosity and permeability for packs of unconsolidated spheres (and by implication for clean, well-rounded sands) is only indirectly applicable to chert-shale mixtures. Hewitt (1966) has shown the general relationships between permeability and the combined content of silt and clay in various North American sandstones but the relationships shown in Figure 7.6 are valid only for the individual stratigraphic units. There is a rough inverse correspondence between measured permeability and gamma ray response within the Osage in the Hart #1 well. Therefore, shale ratios estimated from gamma ray logs can be regarded as imperfect measures of permeability.

Areal Variation of the Mississippian B

The three geologic variables measured in the 124 wells were independently contoured using the SURFACE II computer program described in Chapter 3. Contour grids were generated with a spacing between rows and columns set at one-quarter mile, which is appropriate for the well density in the area. These grids were later used to calculate shape measures of the surfaces, employing some of the techniques described in Chapter 4.

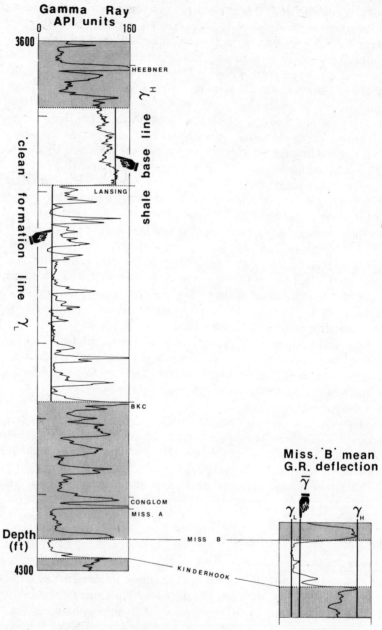

FIGURE 7.5. Method of estimating shale ratio from gamma ray log for Mississippian *B* zone illustrated by log of Western Petroleum Hart #1 well.

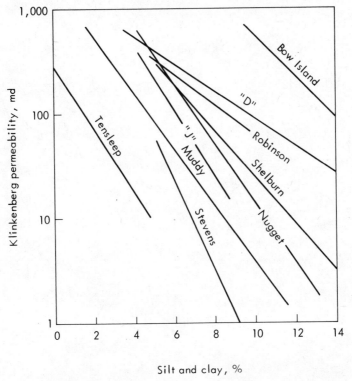

FIGURE 7.6. Generalized relationships between permeability and percentage of silt and clay in various sandstones that serve as reservoir rocks. Modified from Hewitt (1966).

A structural contour map of the top of the Mississippian *B* is shown in Figure 7.7. Structural elements corresponding to regional and local features can be separated by fitting a linear trend surface (Figure 7.8) to the subsurface elevations. The linear trend accounts for 51 percent of the total variance and implies that the surface is composed in equal measures of a simple regional dip and more local structures. The regional trend equated with this linear surface dips about 16 feet per mile to the south-southwest. Deviations from the trend surface (Figure 7.9) form two sets of local features, one whose grain is oriented approximately parallel to the regional dip, and the other to the regional strike. The orientation probably reflects tectonic elements aligned with major structural trends of the Central Kansas Uplift and with the Nemaha Anticline. On a finer scale, local elements may reflect small-scale tectonic deformation, as well as possible modification by ancient weathering processes. It should be noted

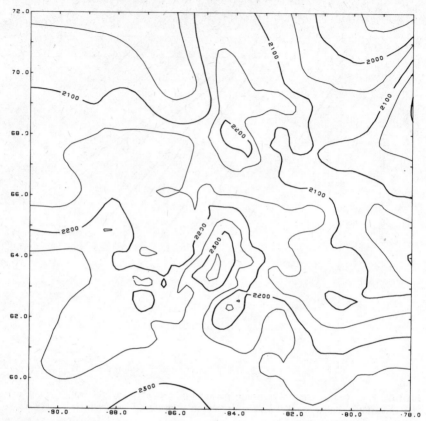

FIGURE 7.7. Structure of top of Mississippian *B* in study area, expressed in feet below sea level.

that part of the area that has been contoured involves extrapolation where the Mississippian *B* has been removed by erosion.

The locations of oil and gas fields show a consistent relationship with the trend surface residuals. While the fields do not coincide with residual "highs," as would be expected if they were strictly controlled by structure, there is a tendency for fields to be aligned along the flanks of local positive structures, where many of the fields occur on the zero-value contour line of the residual map.

Thickness contours of the Mississippian *B* (Figure 7.10) show a pattern of elongate, hummocky ridges of thicker development that extend across the area. Although these thicker trends follow the local structural grain, the axes of maximum thickness are displaced laterally from local struc-

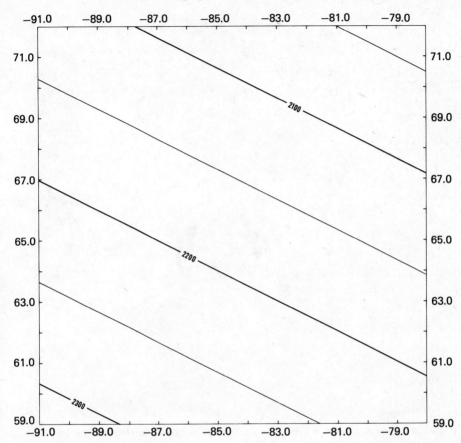

FIGURE 7.8. Linear trend surface of Mississippian *B* structure. Contours in feet below sea level.

tural highs. This suggests that irregular wedges of sediment accumulated along the flanks of local structures and thinned over both adjacent highs and intervening structural lows. Thus, the locations of oil and gas fields are closely associated with the trends of maximum thickness in the producing interval.

There is a general inverse relationship between the thickness and the shale content of the Mississippian *B* (Figure 7.11). As a consequence, the thicker developments that flank local structures are relatively "clean" chert accumulations, in contrast to thinner, more shaly lateral equivalents in adjacent structural highs and lows. These relationships are shown by the block diagram in Figure 7.12, where shale ratios greater than 0.2 are

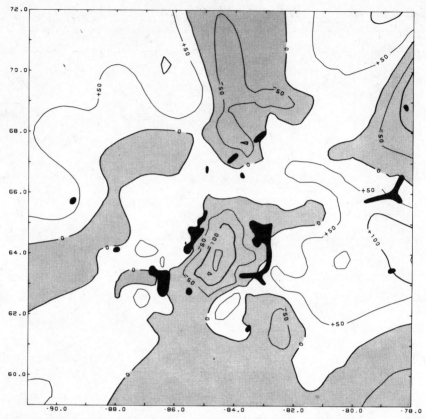

FIGURE 7.9.   Residuals, in feet, from linear trend surface of Mississippian *B* structure. Negative residuals are shaded and positive residuals are unshaded.

shown by shading. Relief on the block diagram represents thickness of the Mississippian *B*. Most of the fields are located within areas where shale content is low, suggesting that local permeability variations were important influences in trapping the oil.

Correlation coefficients were computed between shale ratio, thickness, and structural residuals for all 124 wells (Table 7.2). There is a pronounced negative correlation between shale ratio and thickness but no other significant correlations. The inverse relationship between shale content and thickness has already been noted in reviewing the contour maps of these variables. These maps suggest that the thickest (and least shaly) developments tend to be located along structural flanks. Consequently, significant correlations should not be expected between struc-

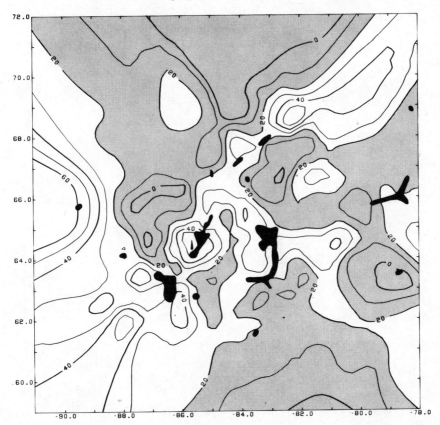

FIGURE 7.10. Isopach map of Mississippian *B* zone. Areas thinner than 20 feet are shaded, whereas areas thicker than 20 feet are unshaded.

tural residuals and thickness, or between structural residuals and shale ratio. Instead, there is a more complex inverse relationship between the absolute values of the structural residuals and thickness of the Mississippian *B*.

The contour maps of the geologic variables suggest that localities of oil reservoirs can be distinguished statistically from intervening dry areas. These differences can be examined by separating the wells into a set of 33 producing wells and a set of 91 dry holes. For each set, the means and standard deviations of the three variables were computed and are shown superimposed on the frequency histograms of Figure 7.13.

In terms of shale content, no producing well penetrates an interval whose shale ratio exceeds 0.3. Furthermore, the mean shale content for

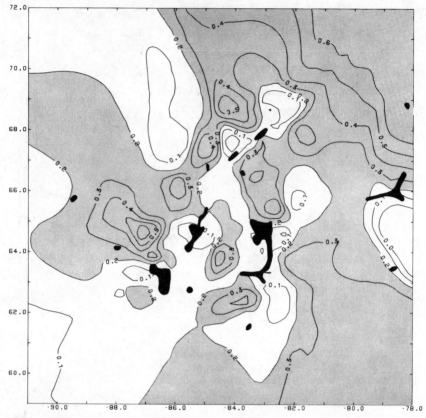

FIGURE 7.11.   Mississippian *B* shale ratio. Areas with ratio greater than 0.2 are shaded, whereas areas with ratio less than 0.2 are unshaded.

the producing wells is clearly distinguished from the mean for the dry holes. Similarly, production appears to be limited to places where the Mississippian *B* is thicker than 15 feet. The mean thickness for the producing wells is markedly greater than that of the dry wells. The contrast between the structural residuals in the two sets of wells is more subtle because the means are similar. However, the standard deviation is significantly smaller for the producing wells, which suggests that the fields tend to be concentrated at the flanks of local positive structures.

Locus and Context Measures

The geologic variables measured in each well pertain only to a single geographic location or *locus*. In other words, measurements in a single

TABLE 7.2. Correlation coefficients between Mississippian B variables based on all wells used.

		A	B	C
A = Shale ratio	A		-0.52^*	0.11
B = Thickness	B			0.02
C = Linear structural residuals	C			

*Significant at 1 percent level.

well record subsurface characteristics at a point and provide no information about how these properties change laterally. However, the accumulation of oil or gas is governed not only by the presence of a suitable reservoir rock, but also by a suitable trap configuration. Therefore, in addition to measuring the reservoir capabilities at a site, we need an estimate of the lateral variations in the geologic variables to ascertain whether there is a trap.

To clarify this distinction, two hypothetical situations are illustrated in Figure 7.14. In the upper example the shale content of a potentially producing interval is contoured across an area. Assume that permeability is inversely related to the shale ratio. Wells drilled at sites A and B encounter a shale ratio of 0.1, which corresponds with good reservoir

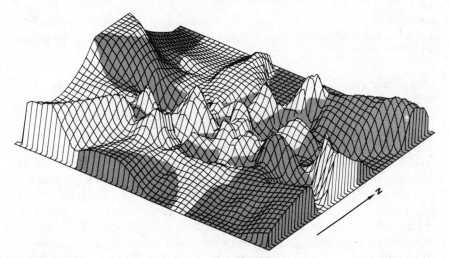

FIGURE 7.12. Mississippian B shale ratio variations superimposed on thickness relief model of study area. Areas with shale ratio greater than 0.2 are shaded. Vertical range of thickness is from 0 to 70 feet. Grid-mesh spacing is ¼ mile. View is from south-southwest at an angle of 15 degrees from horizontal.

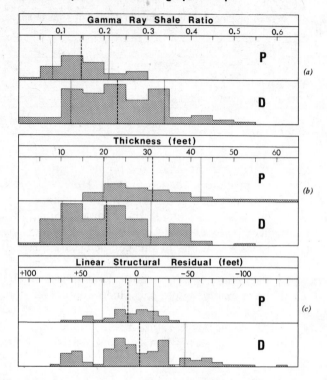

FIGURE 7.13. Frequency histograms of Mississippian B. (a) Shale ratio. (b) Thickness. (c) Structural residuals. Mean values shown by heavy dashed lines. A unit standard deviation is indicated on each side of each mean with dotted lines.

properties. However, well A is dry and well B is a discovery. The different outcomes are due to variations in shale ratio in the immediate vicinity of each well. In the vicinity of well B, lateral variation provides a trap, which is not the case for well A.

In the lower example of Figure 7.14, the contours represent structural residuals from a simple regional trend. Assume that oil entrapment is locally controlled entirely by structure. Both wells C and D are producers, although their structural residual elevations are radically different, being positive for C and negative for D. The reason for the similarity in outcomes is found in the structural configuration of areas immediately surrounding each well. The two wells penetrate extremely similar, closed anticlinal structures which provide suitable traps. These examples illustrate that in evaluating a given location, variables must be defined that treat both the location itself (the "locus measures"), and conditions immediately around the location (the "context measures").

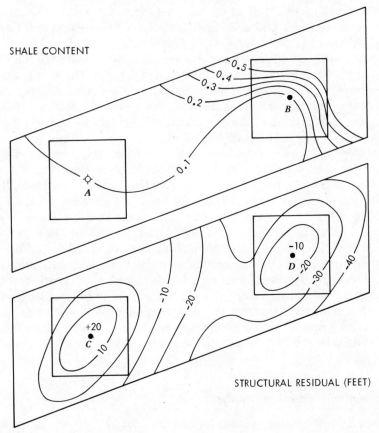

SHALE CONTENT

STRUCTURAL RESIDUAL (FEET)

FIGURE 7.14. Hypothetical examples illustrating contextual relationships of shale ratio and structural residuals.

One method of numerically expressing context measures is given by Demirmen (1973a) who employed polar coordinates to produce sets of numbers that describe the slope and curvature of contoured surfaces. Demirmen's methods (described in Chapter 4) use a series of two-dimensional profiles along polar coordinates that radiate from the center of a square or rectangular segment of area. The procedure is compatible with computer contouring of maps because a grid representing the surface must be calculated before the contours are drawn. The numbers on the grid are supplied along the polar coordinates, and from them a variety of context measures can be computed. From the suite of potentially useful context measures suggested by Demirmen, the "marginal polar slope" was selected for the Stafford County study area. Greysukh (1966) used a similar measure for characterizing land forms. An example of a surface and its marginal polar slope values is shown in Figure 4.2, where a centrally located well serves as the origin for a set of eight polar lines that radiate outward to the edges of a local area of the map. The marginal polar slope associated with a polar line is equal to the difference in elevation at the center and the point along the polar line where it crosses the edge of the local area, divided by the distance from the center to the edge. If eight polar lines are used, a set of eight numbers, each representing the marginal slope along one of the lines, provides the context measure of polar slope for that particular point. Demirmen's polar slope method was applied to contour maps of all three geologic variables in the Stafford County study area. Local areas one mile square were centered at each of the 124 wells. The dimension of one square mile accords generally with the dimensions of gas and oil fields in the study area. Thus, for each of the three geologic variables, nine measures were computed for each well. One measure is a locus measure that pertains to the well's location, and the other eight are context measures that describe lateral variations within the square mile surrounding the well. Since three of the wells were less than half a mile distant from the map margins, definition of one-square-mile areas centered on these wells was precluded. As a result the full set of wells was reduced to 121 for the purposes of locus and context analyses.

Discriminant Function Analysis

The preceding discussion indicates that the location of oil and gas fields in the Stafford County study area can be related to the three geologic variables that were mapped. In a conventional prospect evaluation, the contour maps containing the geologic information would be superimposed and the contribution of each variable would be weighted according to experience and preference. In the Stafford County study, discriminant analysis was used to weight the geological variables according to their effectiveness in distinguishing the locations of producing wells from dry holes. The discriminant analysis computations thus correspond to an "experience factor" of a geologist who has studied the same data, although discriminant analysis is more objective and has the advantage of being quantitative. Discriminant analysis is therefore useful in appraising future drilling sites in terms of their likelihood of success, based on the perceived subsurface geology. This phase of statistical analysis is equivalent to the "prospect evaluation" function of exploration geologists.

Data for the discriminant analysis consist of both the context measures and a locus measure for each of the three geologic variables. Initially the wells were separated into a producing-well set and a dry-hole set. The contributions to discrimination by the context and locus measures were computed and are summarized as percentages in Table 7.3. These measures were used to typify *optimum productive areas*. Listed in order of decreasing contribution to the distinction between producing and dry wells, the six variables are as follows:

1. Lateral change in shale ratio
2. Local thickness
3. Lateral change in thickness
4. Lateral change in structural residual
5. Local shale ratio
6. Local structural residual value

An effective graphic representation of subsurface relationships can be made using areas whose discriminant scores most strongly classify them with the productive population, as indicated by the highest discriminant scores. Figure 7.15 shows "optimum" discriminant score areas which contain simple patterns of variations that are readily interpretable in terms of both reservoir and trapping properties. For example, favorable variation in shale ratio is described by a clean formation in the center of the area, with increasing shale content to the north and east (Figure 7.15a). Favorable thickness variation involves a wedge relationship, with thinning to the east (Figure 7.15b). Favorable structure involves the nose of an anticline that plunges to the northeast, with a flanking structural high to the east (Figure 7.15c). The geometry of these patterns is compatible with regional oil migration in an updip direction to the north-northeast.

Table 7.4 is based on the discriminant analysis and shows that the probability of correct classification of a location as a producing well is 0.90 and that the probability of correct classification of a dry hole location is 0.81. These probabilities suggest that the six geological variables used

TABLE 7.3 Percentage contributions to discrimination made by locus and context measures of Mississippian *B* variables.

	Gamma Ray Shale Ratio	Thickness	Linear Trend Surface Structural Residual
Locus	9.4	23.2	1.1
Context	29.4	18.6	18.3

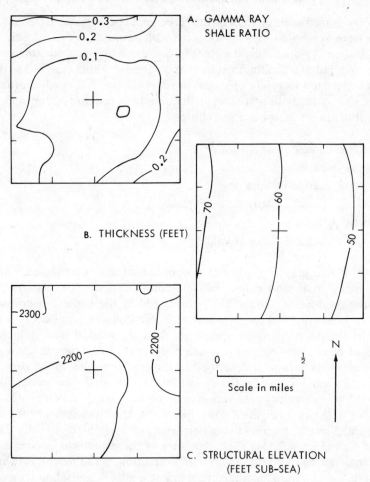

FIGURE 7.15. Submap areas with high partial discriminant scores and therefore classified as optimum locations for oil production in Mississippian *B*.

TABLE 7.4. Discriminant analysis classification of Mississippian *B* well sites into dry and producing categories based on locus and context measures.

Actual Status of Wells	Population 1 (Producer)	Probability	Population 2 (Dry)	Probability
Producing wells	28	.90	3	.10
Dry wells	17	.19	73	.81

are effective in statistically distinguishing productive locations from dry spots in the set of 121 wells.

Discriminant scores were calculated for each well, then posted and contoured (Figure 7.16). The resulting surface can be regarded as a composite measure of the geology that is specifically keyed to oil occurrence. By tracing the contour line which coincides with the discriminant index, the area may be divided into two discriminant-score categories, one containing potentially productive localities, and the other containing unfavorable areas as shown in Figure 7.16. When compared with Figure 7.3, this discriminant score map shows the correspondence between known fields and areas classified as potentially productive. Localities that

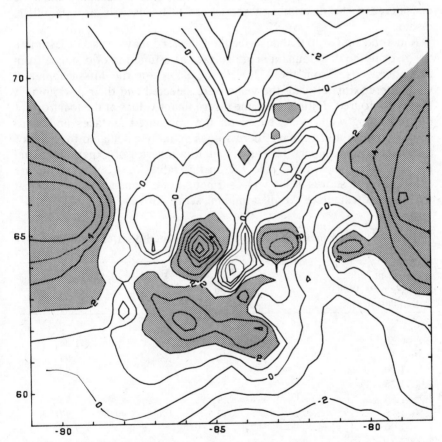

FIGURE 7.16. Discriminant score surface of Mississippian B. Shaded areas are classified as potentially productive.

are in the favorable class, but not yet tested by drilling, should be considered as favorable for future exploration.

Estimation of Error in the Discriminant Score Map

The discriminant score at each well is simply a linear combination of the original geological variables measured in the Mississippian *B*. Since the original variables can be considered as a continuous surface, their combination into a discriminant score is also a continuous surface. Although at each well a discriminant score can be calculated directly, in areas between wells the discriminant value must be interpolated. Just as interpolation of the original variables involves a degree of uncertainty, so does contouring of the discriminant scores; this uncertainty must be considered if the discriminant scores are to be used in a probabilistic fashion.

A method similar to that described in Chapter 5 may be used to estimate the error in the discriminant score surface as a function of distance from available wells. From the 121 wells available for the Mississippian *B* study, approximately half the wells were selected and their discriminant scores contoured. Estimates of the discriminant values at the locations of the remaining wells were based on the contoured surfaces and were compared with discriminant score values actually calculated for each of the wells. This comparison was repeated in a series of experiments and the results are shown in Table 7.5. In this table, the root mean square (RMS) error in scores is shown as a function of distance from the nearest well used to calculate the discriminant score surface.

TABLE 7.5.   Root mean square error of interpolated discriminant scores as a function of distance from nearest well.

Distance in Miles from Nearest Well in Simulation Run	Number of Observations	Root Mean Square Error
<0.25	63	1.14
0.25–0.50	243	1.36
0.50–0.75	116	1.94
0.75–1.00	75	1.85
1.00–1.25	52	1.96
1.25–1.50	19	2.14
1.50–1.75	12	1.77
1.75–2.00	9	2.44
2.00–2.25	6	5.10
2.25–2.50	4	2.09

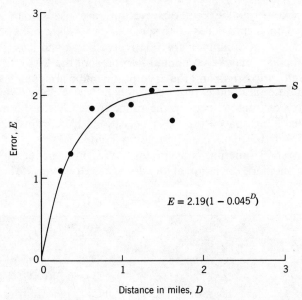

FIGURE 7.17.   Root mean square error of interpolated discriminant score versus distance from nearest well control.

The RMS/distance estimates are plotted in Figure 7.17. Clearly, the function is not the simple linear relationship associated with the error estimation of the Lansing structural surface, described in Chapter 5. Complexity in the RMS function of the Mississippian B may be attributed to several causes. The number of observations used in estimating the error is relatively low, particularly for the greater distances involved. More simulations would increase the sample numbers, but successive simulations are not independent and there would be a high degree of redundancy. Furthermore, the geological variables appear to be anisotropic, reflecting the fact that there are structural and stratigraphic trends aligned with the regional dip. The error function as computed here, however, is isotropic (unrelated to geographic orientation) and represents an average of directional components. The computed error function therefore gives only an approximation of the errors in interpolation which are a function of distance from control well.

In fitting the RMS curve, two assumptions are reasonable. First, the error at the wells themselves can be considered to be zero. Second, the function becomes asymptotic to an upper bound equal to the standard deviation in the discriminant scores for the entire map. Fitting the graph of the RMS function by least squares to the data points is hampered because variable numbers of observations are used in their separate estimation. This would necessitate some kind of weighting procedure that would favor errors computed for relatively short distances. A modified exponential curve may be computed to represent the RMS/distance relationship. Since the origin and the asymptotic value are specified initially, only one additional point is required to characterize the curve. The most reliable error estimates occur in the range from 0.25 to 0.5 miles, since these estimates are based on 243 observations. Accordingly, this value was used to fit the exponential curve on Figure 7.17.

Since the RMS function is asymptotic, the curve never quite reaches the limiting standard deviation at any distance. However, if the maximum

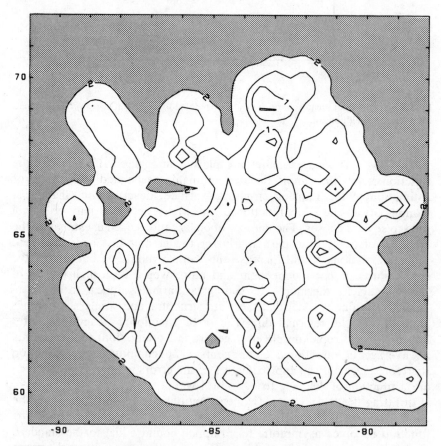

FIGURE 7.18. Uncertainty map of estimated error in interpolated discriminant score values shown in Figure 7.16. Values are in discriminant score units.

extent of autocorrelation of the surface may be informally identified with an error value of 95 percent of the standard deviation, then the range over which surface estimation may be made is approximately 1¼ miles. The distance is less than the corresponding range of the Lansing structural surface described in Chapter 5, because the discriminant scores vary more over a smaller distance. The discriminant score surface reflects the magnitude of the geological features as well as the relationship between discriminant scores and oil occurrence, which in turn is influenced by the areal dimensions of the oil fields and their geographic locations.

The Discriminant Score Uncertainty Map

A grid matrix was computed in which each grid point was tagged with the distance from the nearest well. These distances were transformed to expected errors by using the error function of Figure 7.17, and the resulting uncertainty values were contoured (Figure 7.18). The uncertainty contours circle around the control wells, reflecting the relatively short range in autocorrelation in the discriminant scores. On a regional scale, the northern edge of the contoured area broadly parallels the subcrop margin of the Mississippian B. Because erosional truncation of the B is not incorporated in the discriminant scores, the closely spaced parallel contours reflect the lack of well control in the updip direction. Thus, the gross features of the error map emphasize that prospects can be evaluated with relative certainty only in the central part

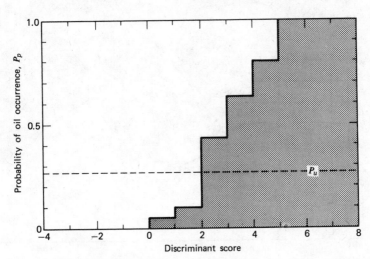

FIGURE 7.19. Histogram of conditional probabilities of oil occurrence (P_p) based on discriminant scores related to unconditional probability (P_u).

of the mapped area. Near the edges of the map the discriminant score surface (Figure 7.16) represents only an extrapolation from available well control.

Probability of Oil Occurrence in the Mississippian *B*

The linear discriminant function provides a maximum "separation" between the set of producing wells and the dry wells. The accompanying discriminant score scale can be used to compute a contingency function relating a discriminant score interval to the probability of oil occurrence, based on producing well and dry hole frequencies (Figure 7.19).

Because the probabilities are based on numerical frequencies of wells

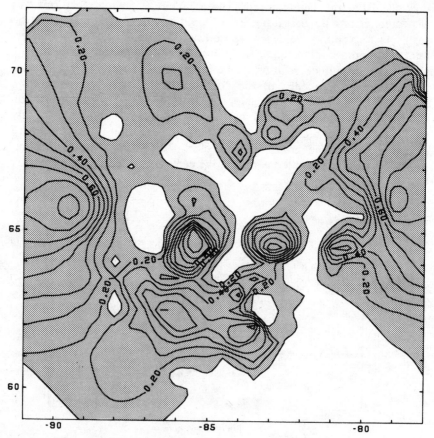

FIGURE 7.20. Probability surface of Mississippian *B* relating oil occurrence to discriminant score variations. Prospective areas are shaded.

rather than areal measures of field size, they are informal estimates which must be judged against a "background" probability. Since the region was at an intermediate stage of maturity when studied, some undiscovered oil fields may be presumed to exist. As a consequence, probability measures based on the relative areas underlain by production are impractical. Because the probabilities are based on well frequencies, the inclusion of production wells in the study sample results in a bias toward probabilities favoring oil occurrence. An unconditional probability, P_u, may be computed by dividing the number of productive wells by the total wells in the sample. P_u represents the probability of oil occurrence which is not based in any way on knowledge of geological variation in the Mississippian B. It constitutes a background level for comparison purposes. Probabilities conditional on discriminant scores reflect the influence of geologic factors and therefore define prospective or dry localities, as related to P_u. The intersection of the conditional probability histogram with the overall unconditional probability level coincides approximately with the discriminant index. Since the discriminant index best discriminates the dry from the producing localities, this coincidence demonstrates the internal consistency of data used in the study.

By combining the contoured discriminant score surface with the estimated error surface, a probability surface can be computed. The resulting map (Figure 7.20) outlines favorable anomalies, some of which coincide with known fields. Favorable areas that do not coincide with known fields should be considered as potential exploration sites. However, exploration decisions based on this map should be tempered by the following considerations: (a) the surface is estimated from a limited collection of control wells; (b) the fitted error function is isotropic and does not consider preferred orientation of geologic variation; (c) there is no information in the data concerning absence of the Mississippian B; and (d) the probability values are relative rather than absolute measures.

CHAPTER 8
Probabilistic Financial Analysis

EXPECTED MONETARY VALUE

The ultimate objective of probability methods in oil exploration is to make better decisions. Effective decision theory applied to exploration requires coordination of the two rather disparate disciplines of geology and economics. Specifically, it involves the linkage between geological evaluation of prospects and economic analysis through numerical estimates of outcome probabilities. Probabilities derived from geological assessments can be incorporated directly into economic decision methods in which alternative financial outcomes are objectively and quantitatively assessed. Grayson (1960), Megill (1971), and Newendorp (1976) provide introductions to these techniques as applied to oil exploration; the principles of the techniques are outlined below.

First, consider the construction of a "payoff table" that lists possible outcomes such as a dry hole or a million-barrel discovery versus possible alternative acts, such as not drilling or drilling with 100 percent working interest. An example of an abbreviated payoff table is shown in Table 8.1. Clearly, this table indicates that the most favorable act, if a two million barrel field were discovered, is to drill the prospect with 100 percent working interest. However, no consideration has been given to the probabilities of the different outcomes. Desirable as a two million barrel discovery is, it is far more probable that a dry hole will actually result from a specific drilling venture.

The next step is to assign probabilities to the possible outcomes as

TABLE 8.1. Hypothetical payoff table listing possible acts and consequences*.

Possible Outcomes in Barrels Discovered	Don't Drill	Drill with 100 Percent Working Interest	Farmout with ⅛ Override
Dry hole	$0	−$50,000	$0
150,000	$0	$425,000	$71,000
500,000	$0	$918,000	$141,000
2,000,000	$0	$1,138,000	$171,000

*Volumes of oil discovered are net after lessor's royalty has been deducted. Producing rate is assumed constant at 15,000 barrels per year. Present price of oil is $6.00 per barrel, and climbs at rate of 0.25 percent a year. Dry-hole cost is $50,000 and producing-well cost is $70,000. Production costs are ignored. Discount rate is 16 percent. Disproportionately small difference in dollar value between half-million and two million barrel discoveries is consequence of low constant production rate coupled with high discount rate. Alternative assumptions, such as production rate proportional to discovery magnitude, would have large effect on financial outcome of given act and given outcome.

illustrated in Table 8.2. The financial consequences for each outcome and act must be multiplied by the probability estimate for that outcome, and the succession of products summed. The sum is the *expected monetary value* (EMV) for that act, and it is this value which is of prime importance in making a decision. Note that in Table 8.2 the expected monetary value of the act of drilling with 100 percent interest is only $16,000, even though the payoff for a two million barrel discovery would be over a million dollars. However, the probability of a two million barrel discovery is very low (0.002, or 1 in 500 in this hypothetical example), whereas the probability of a dry hole is high (0.600).

In calculating the payoff table, an additional refinement involves assumptions about the future price of oil and assigning probability estimates to these prices. If the assumptions are made that (*a*) the probability that oil prices will remain stable is 0.50, (*b*) the probability that oil prices will rise by 2 percent a year is 0.25, and (*c*) the probability that oil prices will decline by 1 percent a year is 0.25, then the expected future price of oil 10 years hence will be $6.16 per barrel if today's price if $6.00, and so on. Estimates of the future value of oil are implicit in every oil exploration decision, whether estimated quantitatively or not.

Note that in Table 8.2 the act with the highest EMV is a farmout, with return after payout of the well's cost with a 50 percent working interest. However, all of the acts listed in the table have positive EMVs, although there is wide range in their value.

TABLE 8.2. Expansion of hypothetical payoff table presented in Table 8.1 with expected monetary values (EMVs) for each act listed in bottom row. Figures are in thousands. For example, act of choosing 100 percent working interest has an EMV of $16,000. Economic and production assumptions are identical to those listed for Table 8.1.

| Possible Outcomes after Deducting Royalty (Millions of Barrels) | Prob- ability | Possible Acts with Payoffs | | | | | | | |
| | | Working Interest (Percent) | | | | Farmout Override | | Farmout, Back after Payout | |
		100	75	50	25	1/8	1/16	25% Working Interest	50% Working Interest
0	.600	−50	−37	−25	−12	0	0	0	0
15	.150	−7	−5	−3	−1	9	4	0	0
30	.100	51	38	25	12	17	8	12	25
45	.070	107	80	53	26	25	12	26	53
75	.040	211	158	105	52	40	20	52	105
150	.020	425	318	212	106	71	35	106	212
300	.010	712	534	356	178	112	56	178	356
500	.005	918	688	459	229	141	70	229	459
1000	.003	1099	824	549	274	166	83	274	549
2000	.002	1138	853	569	283	171	85	284	569
Expected monetary value of each act		16	11	7	3	10	5	11	23

Each EMV at the bottom of Table 8.2 should *not* be considered as a forecast of the specific financial consequences of a particular act. The financial consequences of each alternative act, coupled with each possible outcome, are listed in the body of the table. The EMV for a given act represents a selected weighting of the financial consequences for that act according to the outcome probabilities involved. In other words, the EMV attached to each act can be considered to be a statistical forecast that summarizes both the favorable and the unfavorable financial outcomes in a single number.

THE CONCEPT OF UTILITY

As useful as the payoff table is, it includes no consideration of the investor's risk-taking ability. An act may have a strongly positive EMV, for example, but could expose an investor to a disastrous loss if the dry-hole cost exceeded his available capital. Clearly, some modification of the payoff table is needed to incorporate the investor's risk-taking ability. For example, a small operator's risk-taking ability is generally much more limited than that of a major oil company. The risk policy of a company or an individual is conveniently expressed in terms of the *utility* concept originally advanced in the eighteenth century by Bernoulli, and expanded in modern times by Von Neumann and Morgenstern (1947). Grayson (1960) provides an excellent introduction to the use of utility in oil-exploration decision making, and Newendorp (1967) describes two methods of obtaining utility functions.

The utility concept provides a way to describe the consequences of a given event and act in terms of its desirability. This may be done by developing a function (conveniently represented by a curve) which relates financial outcomes to *utiles*. Utiles are arbitrary units and are a measure, positive or negative, which represents the consequences of a variety of financial outcomes to an investor. Individuals have utility functions, as do organizations. A utility function shows the relationship between an individual's (or company's) desire for gains (positive utility), versus the willingness to suffer losses (negative utility) in the pursuit of the same gains. An individual who is strongly risk averse exhibits a reluctance to sustain losses that is much greater than his corresponding desire to reap gains. Consequently, he will have a utility function that slopes much more steeply in the negative portion than in the positive (as illustrated in Figure 8.1). If this utility function represents that of a particular oil operator faced with the alternatives analyzed in Table 8.2, values in this table can be converted to utiles that are specific to his outlook, and in turn the

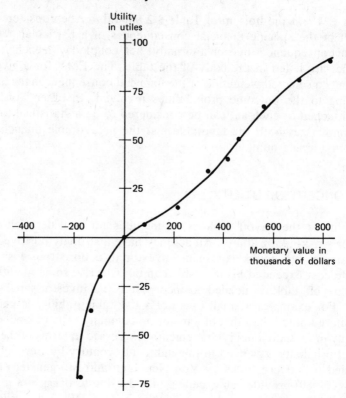

FIGURE 8.1.   Graph of hypothetical utility function used in calculating utility values in Table 8.3. Utility function pertains to individual investor or organization.

expected utility value of each act can be computed. The results are presented in Table 8.3, which is similar to the payoff table. The expected utility value thus provides an objective means of choosing the most appropriate act by selecting the one that has the highest expected utility.

The act with the greatest positive expected utility value is to farm out the prospect and then come back for a 50 percent working interest after payout. In fact, the utility table shows that although all four acts involving direct working interests without farmout provisions have positive EMVs, they all have negative expected utility values because of the operator's aversion to loss, and hence should be avoided by him. Any of the farmout situations are acceptable, however, because all have positive expected utility, although they range widely in their desirability.

For readers unfamiliar with the computation of EMVs and expected utility values, the following example calculation provides an illustration of

TABLE 8.3. Utility table that contains the monetary consequences presented in Table 8.2 converted to utiles through use of the utility function graphed in Figure 8.1.

Possible Outcomes (Millions of Barrels)	Probability	Possible Acts with Payoffs in Utiles							
		Working Interest (Percent)				Farmout Override		Farmout, Back after Payout	
		100	75	50	25	1/8	1/16	25% Working Interest	50% Working Interest
0	.600	−9	−9	−8	−5	0	3	0	0
15	.150	−3	−2	−1	0	0	0	0	0
30	.100	1	1	0	0	0	0	0	0
45	.070	5	3	1	0	0	0	0	1
75	.040	15	9	5	1	1	0	1	5
150	.020	45	30	15	5	2	0	5	15
300	.010	83	61	35	11	5	1	11	35
500	.005	98	81	50	18	8	2	18	50
1000	.003	101	93	63	24	10	3	24	63
2000	.002	102	95	66	25	11	3	25	66
Expected utility of each act		−2.1	−3.3	−3.6	−2.7	0.3	0.1	0.6	1.6

the procedure applied to data drawn from the 100 percent working interest columns of Tables 8.2 and 8.3. In computing the EMV, the probability of each outcome (column A) is multiplied by the payoff or monetary consequence for that outcome (column B) and the resulting products (column C) are summed:

A Probabilities	B Monetary Consequences	C Product of $A \times B$
.600	−$50,000	−$30,000
.150	− $7,000	−$1,050
.100	$51,000	$5,100
.070	$107,000	$7,490
.040	$211,000	$8,440
.020	$425,000	$8,500
.010	$712,000	$7,120
.005	$918,000	$4,590
.003	$1,099,000	$3,297
.002	$1,138,000	$2,276

$$\text{EMV} = \$15,763$$

To compute the expected utility, essentially the same computations are made after transforming each outcome's payoff in dollars to its corresponding utility value. This transformation depends on the utility function of the individual or company and can be made by direct reference to the graph of the appropriate utility function. If the utility function of Figure 8.1 is used, the various monetary consequences of Table 8.2 correspond to the following utility values:

B Monetary Consequences	D Corresponding Utility Values
−$50,000	−9
−$7,000	−3
$51,000	1
$107,000	5
$211,000	15
$425,000	45
$712,000	83
$918,000	98
$1,099,000	101
$1,138,000	102

The expected utility value associated with 100 percent working interest is then derived in a parallel manner to the EMV with substitution of utility values for monetary consequences. By summing the succession of products of probability and utility value for each outcome, the resulting expected utility value expresses the relative desirability or undesirability of the course of action.

A Probabilities	*D* Utility Values	*E* Product of A × D
.600	−9	−5.40
.150	−3	−.45
.100	1	.10
.070	5	.35
.040	15	.60
.020	45	.90
.010	83	.83
.005	98	.49
.003	101	.30
.002	102	.20

Expected utility = −2.08

Derivation of Utility Functions

The validity of the utility concept has been widely accepted by business theorists for several decades and the concept itself was outlined as early as 1738 when Daniel Bernoulli presented a paper on the subject to the Imperial Academy of Sciences in Saint Petersburg (reprinted in 1954 in *Econometrica*). Utility theory recognizes that an economic decision is not determined solely by the expected monetary outcome but also by the risk of capital involved and the assets of the entrepreneur, a concept which is almost axiomatic for the modern businessman.

Although utility is recognized as a conceptual reality, the practical problems of defining an appropriate utility function for an individual, small company, or corporation are so great that its use is limited. Instead, utility theory has been largely applied in an intuitive manner to the interpretation of expected monetary value calculations.

Both Grayson (1960) and Newendorp (1967) attempted to incorporate utility theory into decision making by establishing the utility functions of oil companies from the results of questionnaires that posed realistic although hypothetical exploration decisions. Their experiments sought to establish the *indifference point* for each decision. This was done either by isolating the probability of success which the company viewed as the critical lower limit in each case, or by probing for the minimum payoff that would make the proposition attractive to the respondent. By plotting the values from the questionnaire, they were able to plot points on utility functions which described the respondent's exploration posture. Not surprisingly, different utility functions were obtained for officers within the same company, and capricious shifts in utility function occurred from time to time for a single individual.

As an alternative, Newendorp (1968) suggested that utility functions describing individual companies should be based on analyses of actual decisions made by the company in real exploration ventures. This has the distinct advantage of gauging the real risk attitudes of the company when faced with decisions involving actual money. The true performance in such circumstances may be quite different from the company's own interpretation of its decision behavior when tested with hypothetical proposals.

Smith (1972) proposed an interesting alternative by defining utility functions with a few key parameters in their formulation. Rather than attempting to fit a utility function to the results of decisions made by the company or individual, the function can be derived mathematically from numerical characterizations of the risk posture of the company ("risk-averse," "risk-indifferent," and "risk-seeking") and the monetary outcome ranges to which these postures apply. Simple examples of pure risk-averse, risk-indifferent, and risk-seeking utility functions are shown in Figures 8.2 to 8.4. These may be compared with the more general type of utility function that incorporates both risk-averse and risk-seeking components, as illustrated by Figure 8.1. The particular example of a general utility function in Figure 8.1 includes a lower monetary limit that corresponds to the company's critical value of loss where the utility function becomes infinitely negative.

Any utility function has certain basic properties that relate utility (u) with the monetary value (x). These properties are as follows:

1. The utility function is monotonic, so increasing monetary values are matched by increasing utility values. The slope of the utility function is always positive and may be expressed in terms of calculus as

$$\frac{du}{dx} > 0$$

where du/dx is the first derivative of the utility function.

2. The risk attitude of any segment of a utility function ("risk-averse," "risk-indifferent," or "risk-seeking") is determined by the sign of the second derivative of the utility function, d^2u/dx^2 in that segment:

Risk-averse: $\dfrac{d^2u}{dx^2} < 0$

Risk-indifferent: $\dfrac{d^2u}{dx^2} = 0$

Risk-seeking: $\dfrac{d^2u}{dx^2} > 0$

These relationships are apparent from Figures 8.2 to 8.4 and simply mean that in the risk-averse case, the rate of change in utility slows with increasing monetary value; that the risk-indifferent case assumes a constant relation between monetary value and utility; and finally in the risk-seeking case, that the rate of change in utility rises as monetary value increases.

3. Most realistic utility functions have a risk-averse aspect. They are bounded by an upper limit since, in general, increasing additions in value are not matched by correspondingly large increases in utility.

4. Utility functions are bounded on the left by a critical value of monetary loss where the utility function becomes risk-averse and drops rapidly toward an infinitely negative utility value.

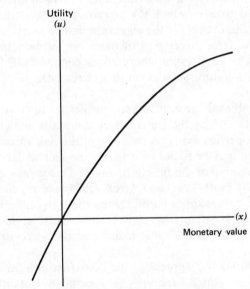

FIGURE 8.2. Generalized risk-averse utility function ($d^2u/dx^2 < 0$). Positive utility is plotted at top, positive monetary value is toward right.

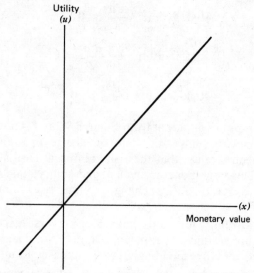

FIGURE 8.3.   Generalized risk-indifferent utility function ($d^2u/dx^2 = 0$).

5. A risk-indifferent utility function lies at the neutral boundary separating risk-seeking from risk-averse functions. A risk-indifferent utility function is identical to a purely EMV approach in that losses are treated identically to gain, except for the algebraic sign. While large organizations may be risk-indifferent over part of the range of monetary value, there must inevitably be some point where losses become sufficiently large that a posture of risk-indifference is no longer realistic.

Smith (1972) used these fundamental properties to construct "parametric utility functions" as the outcome of the mathematical relationships which these properties imply. A measure of the risk attitude expressed in a utility function may be found by dividing the second derivative, d^2u/dx^2, by the first derivative of the function, du/dx. This measure was proposed independently by Pratt (1964) and Arrow (1971) and is generally known as the *absolute risk aversion function*. Since the sign of the first derivative, du/dx, is always positive, the sign of the risk aversion function will describe risk attitude in the same manner as the second derivative, d^2u/dx^2.

Using this parametric approach, the construction of a utility function such as shown in Figure 8.1 requires the specification of only a few critical values. The function illustrated is composed of a risk-averse segment in the area of negative return, linked to a risk-seeking curve at the origin

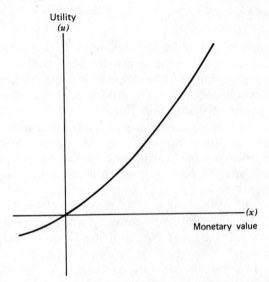

Utility
(*u*)

(*x*)
Monetary value

FIGURE 8.4.   Generalized risk-seeking utility function ($d^2u/dx^2 > 0$).

which becomes risk-averse at higher positive values. By considering each segment separately, the complete curve may be generated. The values required for this operation involve the limits of each segment measured on the monetary value axis, which include (*a*) the critical loss value, (*b*) the origin, and (*c*) the point of inflection linking the positive risk-seeking and risk-averse segments. Also required are the absolute risk aversion functions of each segment, which simultaneously describe the intensity of the risk attitude and the graphical curvature of the function. The complete function is fitted by solving a set of simultaneous equations involving these critical parameters; the reader is referred to the more detailed account by Smith (1972) who derived much of the theory in this area.

The study of parametric utility functions is promising as a rigorous means to specify and analyze the performance of different risk attitudes under various economic constraints and exploration environments. Such research necessarily must be conducted "in-house" by exploration companies in order to gauge the results against the economic realities of everyday operations. Obviously, the findings could be used in several ways by different companies depending on their assets and general exploration philosophy. As a consequence, it is impossible to recommend specific utility functions within the context of this book. However, the importance of defining a utility function as a vital component in exploration analysis must be stressed.

DECISION TREES

Payoff tables and utility tables are examples of how probabilities associated with exploration activities can be used in economic analysis. However, real exploration decisions are part of a long train of decisions in which a particular choice, such as the drilling of a wildcat well, is but one in a series of interrelated acts.

Methods are available to handle more complex decisions, as for example, where there are branching decision paths which form alternative routes that depend on intermediate outcomes. Situations requiring sequences of decisions can be handled with decision-tree analysis, which has been developed to an advanced state by specialists in business decision theory (Coyle, 1972; Raiffa, 1970; Schlaifer, 1961; Wagner, 1969). Figure 8.5 provides a hypothetical example of a decision tree consisting of a series of forks and branches. Some of the forks are "chance forks" (denoted by squares), from which the branch to be followed is a matter of chance, dictated by the outcome of a particular act. The other forks are "decision forks" (denoted by circles), at which the particular branch to be followed involves a decision. Note that the tree incorporates probabilities at a number of points. By working back along the branches of the tree, the EMV at each chance fork can be calculated. As in Table 8.2, the procedure is one of multiplying probabilities by monetary consequences and summing to obtain each EMV. As the decision maker moves from the tips of the branches back toward the origin of the tree, he compares the EMVs of alternative paths at the various forks or junctions. If he chooses the path with the highest EMV at each decision fork, when he eventually reaches the origin he will have selected a succession of acts that yields the highest EMV.

For example, if the operator were at junction E in Figure 8.5, he would see that the act "drill" has an EMV of $136,000 and the act "don't drill" has an EMV of $0. Since he should pick the act with the highest EMV, he would elect to drill. If the operator were at fork O, however, he would see that the act "drill" has an EMV of −$14,000, and "don't drill" has an EMV of $0. The act of drilling would be avoided and the EMV for this fork is $0. The EMVs at the other forks are similarly determined. The rule is to make the decision at each fork that yields the algebraically highest EMV. Thus, even though the algebraically highest EMV may be negative, it still would be the one adopted, and the other path or paths are closed off.

At this point we should note the effects of the two kinds of fork junctions. Point A is a decision fork where the operator has a choice of two alternative branches. He can follow either the "buy seismic" or the

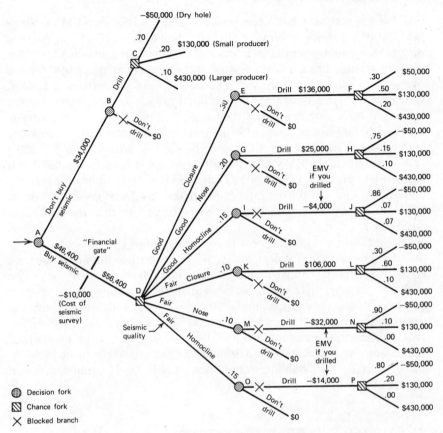

FIGURE 8.5. Hypothetical decision tree. After Grayson (1960).

"don't buy seismic" branch. The buy seismic path leads initially to chance fork D and the route from fork D then depends on the results of the seismic survey. Probabilities attached to the seismic survey's outcomes are shown in the diagram. They necessarily sum to 1.0 because they involve all possible outcomes, consisting of six mutually exclusive results that depend jointly on the quality of the seismic survey ("good" or "fair") and the kind of structure encountered ("closure," "nose," or "homocline"). Thus, chance junction D leads to the six decision forks labeled E through O. In turn, each "drill" branch leads to a chance fork which represents the drilling outcome.

If we work back to the origin along the buy seismic branch, we must multiply the EMVs calculated for the tips of the six subbranches by the probabilities attached to the seismic outcomes. In turn we obtain a single

EMV of $56,400, which is a composite of the six subbranch EMVs. Since the cost of a seismic survey is $10,000 (shown as −$10,000 on the tree diagram), we subtract $10,000 from the $56,400 and the adjusted EMV for the buy seismic branch is thus $46,400. In arriving at this EMV for the overall buy seismic branch, we have combined a succession of EMVs, which in all cases are obtained by multiplying each financial consequence by its attached probability and then summing. Even the seismic survey cost of $10,000 can be thought of in EMV terms since there is absolute certainty (or probability of 1.00) that the seismic cost will be $10,000.

If we calculate the EMV of the don't buy seismic branch, we find that it has an overall EMV of $34,000. Thus, the buy seismic branch has the highest EMV, and if the operator were to base his decisions strictly on an EMV risk policy, he would adopt the buy seismic branch as the preferable alternative.

The question arises as to how these probabilities can be estimated for different junctions in real decision trees. The published literature does not cover this point, possibly because the necessary data are proprietary. The probabilities obviously could be estimated by tabulating frequencies in sequences of events; for example, the success of outcomes of drilling ventures that were preceded by various seismic surveys. Establishing the probabilities would require the careful analysis of a large number of successions of events that led to drilling decisions, but there is no intrinsic reason why such probabilities could not be objectively estimated, given access to appropriate information.

INTEGRATED EXPLORATION SYSTEMS

At this point we have a sufficiently broad overview to appreciate the possibility of developing a fully integrated, analytical exploration decision system. Such a system would necessarily treat both geological and economic information, and would have as its objective the attainment of optimum decisions given a particular explorationist's financial goals and risk position. Figure 8.6 provides a simplified organization chart for such a system. The treatment of geological, geophysical, and production data is represented on one side of the diagram, and business information is represented on the other. The linkage between the two sides is provided by outcome probability estimates.

The oil industry actually uses a less formal procedure that incorporates the main elements identified in Figure 8.6. Geological, geophysical, and production information is maintained in various kinds of files, sometimes accessed by computer. Information is selectively searched for and ex-

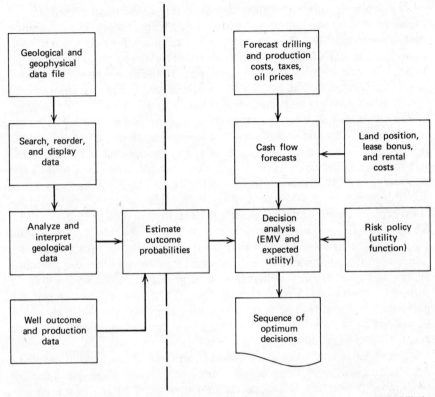

FIGURE 8.6. Simplified organization chart showing main components in probabilistic exploration decision-making system.

tracted from the files and is interpreted. The interpretations are evaluated and exploration prospects are appraised as to their probable outcomes. From the estimates of various outcomes, financial analyses are made, often in the form of cash-flow forecasts. Finally, decisions are reached, with the intention that each is optimum in light of the operator's objectives.

Every one of these steps involves uncertainty, and is therefore probabilistic by definition. Our thesis is that formalized, analytical, mathematically based methods can be used in parallel with the established traditional methods. By the use of mathematically based systems the human decision-maker is not replaced, but instead he is assisted. Mathematical tools implemented on computers have the advantage that many factors can be weighed simultaneously. Furthermore, the decision-maker can consistently and objectively examine a large number of alternatives.

Developing a fully integrated, computer-based decision system for oil exploration is a very large task. To our knowledge, no one has developed such a system, although most major oil companies have developed components. The KOX (Kansas Oil Exploration) System is perhaps one of the most advanced systems at the present time. While by no means complete, nor an integrated whole, KOX does incorporate most of the components which are listed in somewhat greater detail in Figure 8.7.

The elements in the boxes in the lefthand column of Figure 8.7 list some of the capabilities and techniques that have been described earlier in this book. Most of these elements are currently operational although some will undergo continued development. For example, an extensive data file containing well information from Stafford County, Kansas, and from a six-county area in northwestern Kansas has been searched and manipulated repeatedly in preparing the maps described in earlier chapters. The SURFACE II computer graphics system, described in Chapter 3, has been used to manipulate and map subsurface data in these localities. Other computer programs have been employed to calculate discriminant scores and error estimates, and have also provided output which was, in turn, fed into the SURFACE II program for display as contour maps and perspective diagrams. The point is that all components of the system can be used in close coordination with each other, and information flows back and forth between them.

Figures 8.6 and 8.7 emphasize that the principal connection between the "geological" and "business" sides of the system are provided by outcome probabilities. This implies that geological activities are fundamentally directed toward the estimation of probabilities. In turn, analytical tools on the business side require these probabilities for their application. As outlined earlier, the techniques of modern decision analysis such as EMV payoff tables, expected utility tables, and decision trees, are based on such probabilities. Thus, the design of an integrated system must provide for harmonious interdependence between the various components and subcomponents.

PROBABILITY, MONETARY, EMV, AND EXPECTED UTILITY MAPS

Geologists and exploration managers appreciate the usefulness of contour maps for the display of various types of information, including geological features such as structure and facies variation. Furthermore, contour maps can be used to represent statistical measures applied to geological data, such as the estimation error surfaces described in Chapter 5, or

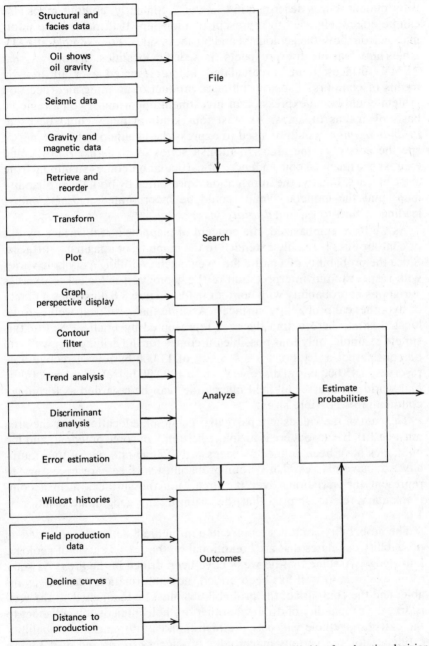

FIGURE 8.7. More detailed organization chart of "geological" side of exploration decision system organization chart shown in Figure 8.6.

discriminant scores described in Chapter 7. Similarly, contour maps also can be effectively used to represent relationships that incorporate information from both the geological and business sides. For example, if EMV tables are feasible for prospects at specific localities, it follows that "EMV surfaces" can be calculated and represented over an area by means of contours. Thus, an ultimate product of an integrated decision system could be an expression of investment opportunities on a regional basis by means of maps. At least four kinds of maps could be used. *Probability maps* could be used to express the likelihood of outcomes of specific acts over the area. In turn, a series of *monetary maps* could express the financial consequences of particular outcomes stemming from a specific act. Finally, the information represented by both the probability maps and the monetary maps could be incorporated in *EMV maps,* leading in turn to *expected utility maps.*

As we have emphasized, the concept of mapping a real surface by use of contour lines is readily extended to the mapping of imaginary surfaces. Since the probability of a particular event such as drilling a dry hole varies with respect to our interpretation of the geology, it is logical to express variations in probability with contour maps. Figure 8.8 illustrates a family of hypothetical probability surfaces. Assume that a wildcat well is to be located somewhere within the area represented by Figure 8.8. For this simple example, only four possible outcomes for the drilling of a well will be considered: a dry hole, a discovery of 13,000 barrels of recoverable reserves, a 45,000-barrel discovery, and a 100,000-barrel discovery. In the real world, of course, oil field magnitudes can be regarded as forming a continuum of possible sizes.

The sum of the values at a particular geographic location over the area must be 1.0, because there is absolute certainty of some outcome, and the outcomes have been defined as necessarily falling into one of four mutually exclusive classes. Thus, four probability surfaces are necessary to represent the variations over the area, and the surfaces form a complementary relationship so that their sum at any geographic location is 1.0.

The probability contours represented in Figure 8.8 must be regarded as probability estimates that are conditional on the present state of geologic knowledge. Assume that no wells have been drilled in the area. As soon as an exploration well has been drilled, new information becomes available, and the contours of the probabilities must be readjusted. If the well is dry, the probability of drilling another dry hole immediately adjacent to the existing dry hole will be quite high (almost 1.0), and the probability estimates for the various magnitudes of success (*b, c,* and *d* of Figure

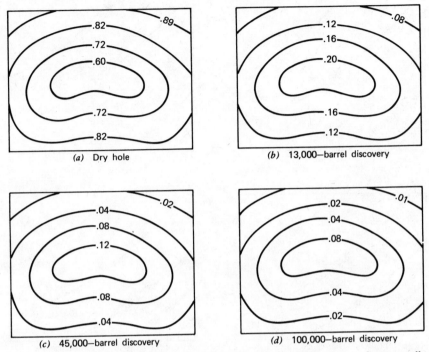

FIGURE 8.8. Series of four hypothetical probability maps which pertain to four mutually exclusive outcomes of wildcat well.

8.8) would necessarily have to be adjusted so that they are low in the immediate vicinity of the dry hole.

For each particular act and outcome, a "monetary surface" could be calculated and represented by contours of dollars over the area. For example, the outcome of a discovery would involve different monetary consequences for the act of drilling with 100 percent working interest, as compared with the act of having the prospect drilled as a farmout by someone else. By treating the various economic aspects, the dollar consequences over the area can be calculated. These may vary depending on differences in drilling and producing costs from place to place. For example, assume that the regional dip is toward the southwest, causing the potential producing horizons to become steadily deeper in that direction. The cost of a dry hole (Figure 8.9) in the northeastern corner of the area is only about $25,000, but is about $135,000 in the southwestern corner. These differences will, of course, affect the dollar consequences in the

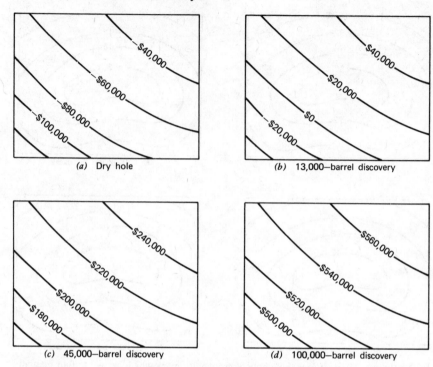

FIGURE 8.9. Series of maps of hypothetical area showing four different monetary outcomes for act of drilling single wildcat well with 100 percent working interest. Outcomes are mutually exclusive and represent all possible outcomes. Value of oil produced after deduction of royalty is $6.00 per barrel. Discount rate is ignored.

event a discovery is made. Discoveries in the northeastern part of the area will be more profitable (or involve smaller losses) than those in the southwest because of lower drilling costs. The monetary values also incorporate the effect of the discount rate. The large discoveries are assumed to take longer to be produced, and therefore some of the oil to be produced in the comparatively distant future has a smaller dollar value when discounted to the present.

Monetary surfaces, as illustrated by the hypothetical examples of Figure 8.9, do *not* involve probabilities. Each monetary surface pertains to a specific act and a specific outcome. An *expected monetary value surface,* however, would involve the probabilistic representation of a series of possible outcomes resulting from a specific act. The value at each point on an·EMV surface would be derived from the probability estimates and the corresponding monetary consequences for all possible outcomes that stem from that particular act. Calculation of an EMV for each point thus involves multiplying a succession of probabilities by their monetary consequences and summing. Figure 8.10a provides an example of an EMV surface that represents drilling with 100 percent working interest. It has been calculated by combining the information from Figures 8.8 and 8.9 at a succession of points over the area. EMV surfaces for other possible acts could be computed in a similar manner.

The final step would be to produce maps of expected utility for each particular act. Figure 8.10b illustrates an expected utility map for the act of drilling with 100 percent working interest, for the hypothetical risk-averse utility function shown in Figure 8.11. The usefulness of an expected utility map lies in the fact that it brings together virtually all relevant aspects of decision making. Geology is considered in the probabilities; exploration costs are considered in the expenses; gains from oil produced (if discovered) are included, and involve producing costs, royalties, taxes, a forecast of future oil prices, and a discount factor; and finally, the operator's willingness to take risks versus his desire for gains are incorporated via his utility function.

Selection of the optimum investment in the hypothetical area of Figures 8.8 to 8.10 is a matter of finding the largest expected utility value. This requires selection of a particular act to be taken at a particular location. If the desirable act at that location cannot be consummated (perhaps the land is already leased), the operator should then consider the location with the next highest expected utility, and so on. All acts considered should have positive utility.

CONCLUDING REMARKS

The material covered in this book is only a beginning; it describes the first exploratory steps in a scientific modus that we believe will be a valuable, even an essential, aspect of petroleum exploration. The next decade will see major changes in the use of geological and geophysical information, including a major shift toward the objective and analytical estimation of exploration outcome probabilities. Within this general forecast, the following areas seem especially deserving of close attention.

1. There is a major need to estimate probabilities attached to specific steps in a decision sequence or chain of events. In any particular oil play, one specific action should not be isolated from the sequence of other events and actions. At any time, the probabilities of future outcomes are dependent on events that have already transpired; the challenge is to

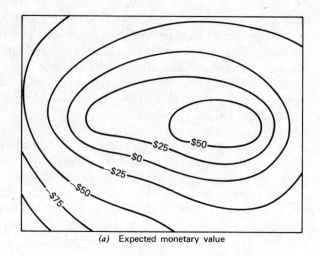

(a) Expected monetary value

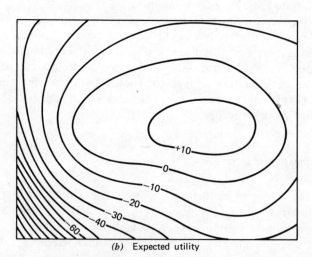

(b) Expected utility

FIGURE 8.10. (*a*) EMV map in thousands of dollars for act of drilling with 100 percent working interest. (*b*) Expected utility map of same area. Contours are in utiles. EMVs in map (*a*) have been transformed to utiles with utility function of Figure 8.11.

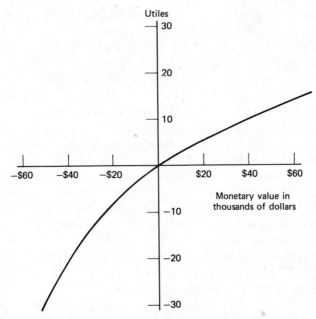

FIGURE 8.11. Utility function used in transforming EMV surface of Figure 8.10a to expected utility surface of Figure 8.10b.

estimate these probabilities attached to a variety of alternative actions. A specific decision can be likened to a particular limb of a large and complex decision tree that bifurcates and branches without end. For practical reasons we must restrict ourselves to consideration of a segment of the overall tree. But, if methods can be developed for estimating probabilities over some reasonable expanse of a branching sequence, it should result in a notable increase in the efficiency of exploration decision making.

2. Outcome probabilities must be estimated in ways that are sensitive to an operator's land position. For example, if an operator has a 640-acre lease, he is not concerned about the probabilities attached to a discovery that involves a field extending over thousands of acres. Instead, he is directly concerned only with the probabilities as they affect his lease, and not those of his neighbors. Adapting probability estimates to specific lease block sizes (and shapes) poses some thorny problems that have received only scant attention.

3. Technological advances have a compounding effect on the difficulties in estimating outcome probabilities. Compare, for example, the effectiveness of seismic techniques used in the 1950s with those presently in use.

Modern methods provide a vastly superior perception of subsurface structures and stratigraphic relationships, sometimes even directly indicating the presence of oil and gas. While these technological advances reduce the uncertainty associated with perception of subsurface geology, they complicate attempts to use historical data in obtaining probability estimates.

4. We have stressed the estimation of probabilities that are conditional on subsurface geology. Yet, there may be major influences on probabilistic estimates that are totally apart from direct geologic and geophysical considerations. For example, the announcement of a wildcat well in a relatively unexplored area will spark renewed interest in the region. If the well is to be drilled by a knowledgable operator, his competitors may assume that the area is geologically favorable even though they have no more direct geological and geophysical information than before the well's location was announced. Thus, there is a need for methods of objectively modifying probabilities on the basis of such indirect information.

5. The economic use of numerical probabilities in exploration decision making needs further development. We have outlined the use of decision tables and decision trees earlier in this chapter, but few oil operators actually use these tools. Perhaps alternatives can be developed that are more familiar and useful to oil operators in everyday exploration decisions.

6. Ways to establish an optimum risk position are needed. Every company and independent operator has a utility function, regardless of whether the company or operator makes conscious use of the utility concept. Quantifying the existing utility function of a company or operator is a major challenge in itself. Objectively defining an optimal utility function, custom tailored for a particular company or operator, is an even greater challenge.

This catalog of as-of-yet unresolved problems most certainly exceeds the list of answers we have suggested in this book. The research reported in the preceding pages has been exciting and rewarding; we look forward to the pursuit of some of these alternative topics in the future. We hope our efforts will encourage others, and that part of the vast scientific, intellectual, and financial resources of the petroleum industry will be directed to these problems. The answers promise to be highly rewarding, both intellectually and financially.

Glossary

Algorithm—A defined operational procedure that leads to a solution of a particular class of mathematical problems.

Ambient probability—The "background" or unconditional probability of success that would be experienced if wells were drilled at random.

Anisotropic—Possessing a "grain" or trend that varies with direction.

Autocorrelation—Measure of the degree of relationship (correlation) between values at a set of points and those at another set of points located a specified distance and direction away.

"Background" probability—See Ambient probability.

Bayes' theorem—A mathematical statement in which prior knowledge of a condition or conditions affects the value of a probability assigned to an event.

Bias—A systematic distortion of a statistical result, as distinct from a random error which is symmetrically dispersed around the result and so balances out on the average.

Bicubic spline—A surface which interpolates from four corner points and which exhibits continuity with adjacent surfaces.

Bimodal frequency distribution—A frequency distribution that when graphed exhibits two distinct peaks.

Binomial distribution—The distribution of the number of successes in n trials when the probability of success remains constant from trial to trial and the trials are independent.

Bivariate—Characterized by two variables.

Canonical variables—A series of variables that are independent and uncorrelated. The matrix of correlations contains zeros except along the main diagonal.

Conditional probability—The probability that an event will occur given that another event has or will occur; for example, the probability that oil will be found, given an anticline is drilled.

Confidence envelope—An interval for which it can be stated with specified probability (called the degree of confidence) that it will contain the parameter, line, or surface it is intended to estimate.

Context measure—An expression of the geometric configuration of a variable around or in the neighborhood of a point.

Contingency table—A table in which individuals or items are classified according to two criteria, one of which forms the columns of the table, and the other forms the rows.

Correlation coefficient—A measure of the linear relationship between two variables, obtained as the ratio of their covariance to the product of their standard deviations. The correlation coefficient ranges from $+1$ (perfect direct relationship) through 0 (no relationship) to -1 (perfect inverse relationship).

Covariance—A measure of the joint variation of two variables around their common (bivariate) mean.

CPU time—Time utilized by the central processing unit of a computer in performing computation or other operations.

Cubic trend surface—See Third-degree trend surface.

Cumulative distribution—A frequency distribution showing how many items are "less than" or "more than" given values.

Decision tree—A diagram that branches to represent alternative courses of action.

Density function—A distribution whose integral (area under the curve) from a to b gives the probability that a corresponding random variable assumes a value within the interval a to b.

Deterministic—A process in which future states can be predicted exactly from knowledge of the present state and the rules governing the process. It contains no random or uncertain components.

Deviation—The difference between an expected or predicted value and the value actually observed.

Discriminant function analysis—A statistical procedure which defines a linear combination of variables that best distinguishes between two or more specified groups. The linear transformation can then be used to classify additional samples into the groups.

Discriminant index—A value midway between the multivariate centers of two groups, used as a decision value for classifying observations.

Discriminant score—The result of transforming multivariate observations by the linear equation found by discriminant analysis.

Domains—The subdivisions of a multivariate spatial representation that can be linked with specific groups of observations.

Drift—A slowly varying component of a mapped variable, analogous to a local average. It is used in kriging as an alternative to trend surface analysis.

EMV—See Expected monetary value.

Engel model—A theoretical set of procedures used in a formalized search strategy.

Error analysis—Evaluation of the differences between predicted values in a map and true values obtained by subsequent drilling.

Error function—A probability distribution that describes the expected variation in estimates of a mapped surface, as a function of distance from control.

Error map—Map of the expected variation in estimates of a mapped surface.

Estimation error—A measure of the variation expected if repeated estimates were made of a parameter or surface by randomly selecting samples from a population.

Expected monetary value—The probability of an outcome multiplied by the dollar consequences of that outcome.

Expected value—Long-term average of a variable.

Factor analysis—A multivariate procedure that extracts the underlying (usually independent) "factors" that give rise to intercorrelations between observed variables. The factors represent basic properties which may not be directly measurable.

Filtering—Isolation of spatial components of specified size or orientation within maps. Filtering is performed by cross-multiplying successive elements of a gridded map with a small filter matrix.

First-degree trend surface—A trend surface consisting of a simple plane, generated by a three-term equation including a constant term and two slope coefficients.

Form-line map—Structural surface represented by lines of constant but undefined values.

Fourth-degree trend surface—A trend surface generated by a polynomial equation having powers up to X^4 and Y^4, and their cross-products.

F-value—Value of a probability distribution for testing the equality of two variances.

Gaussian distribution—See Normal distribution.

Geometric mean—The average of the logarithms of the data values, transformed back to the original units of measurement.

Global fit—A mapping procedure in which a surface (represented by a single mathematical equation) is fitted to all data points simultaneously, as in trend-surface analysis.

Global function—A mathematical function that is fitted to all data points of a map simultaneously.

Histogram—Bar graph in which the area of a bar is proportional to the number of observations within the class limits of the bar resulting from the subdivision of a measured variable on a horizontal axis.

Indifference point—The point at which the acceptance or rejection of a financial opportunity is of no consequence to a decision maker.

Inferential statistics—Those aspects of statistics dealing with the testing of hypotheses, or the making of inferences on the basis of samples.

Isotropic—A property or surface which has no preferred orientation or spatial "grain."

Joint probability—A probability that depends on two or more conditions, as for example, the probability that a particular formation will be encountered that is both highly permeable and structurally high.

Kriging—A contour mapping procedure based on regionalized variable theory. Estimates of a surface are based on a combination of neighboring control points, weighted according to their semivariance.

Law of large numbers—The larger the sample size, the more probable it is that the sample mean comes arbitrarily close to the population mean.

Least squares—A fitting procedure that minimizes the sum of the squared deviations between a fitted equation and the observations.

Level of indifference—The point at which the difference between two (or more) possible financial outcomes is of no consequence to a decision-maker.

Likelihood—The degree of confidence with which an unknown sample may be assigned to one or other of already specified populations.

Linear discriminant function—See Discriminant function analysis.

Linear transformation—A proportional change in a variable, achieved by multiplying or dividing by a constant.

Linear trend surface—See First-degree trend surface.

Local fit—A mapping procedure in which the surface at a point on a map is estimated only from control values within a local neighborhood around the point. More distant control values have no influence.

Local function—A mathematical function fitted to a limited subdivision of a mapped area.

Local residuals—Small areas of residuals or deviations from a trend surface, in which all the residuals have the same sign.

Locus measure—The value of a variable at a point location (see Context measure).

Lognormal distribution—A variable in which the logarithms of the values are normally distributed.

Log transformation—Use of the logarithms of numerical data instead of the original numbers.

Mahalanobis' generalized distance—The square of the straight-line distance between the multivariate centers of two groups of observations. It is measured along the discriminant function.

Marginal polar slope—The slope along a line connecting the center of a rectangular area with a point on the margin.

Marginal probability—A probability value that depends only on a single condition where one or more other conditions exist.

Mean—Average or central value, calculated as the sum of the observations divided by their number.

Model—A geologic hypothesis expressed in mathematical form. A statement of hypothesized relationships between variables.

Monte Carlo simulation—Methods of approximating solutions of problems by sampling from simulated random processes.

Multiple linear regression—A regression technique in which a dependent variable is related to a linear combination of several independent variables.

Multivariate analysis—Statistical procedures in which variations in several variables are considered simultaneously.

Negative binomial distribution—A clustered distribution in which centers of clusters of points are randomly located. The points are scattered logarithmically around the centers.

Normal distribution—A continuous bell-shaped probability distribution specified by two parameters, the mean and standard deviation.

Numerical integration—A method of integrating a function by repeated summation of small increments.

Parabolic trend surface—See Second-degree trend surface.

Parametric utility function—A utility function defined in terms of the slope (first derivative) and rate of change of slope (curvature or second derivative).

Payoff table—A systematic tabulation of the monetary consequences of various acts and various outcomes.

Petroleum resource base—The ultimate amount of recoverable petroleum present, Q.

Poisson distribution—A discrete distribution that describes the occurrence of rare random events.

Polynomial—A linear equation composed of successive powers and cross-products of the variables as in trend surface analysis.

Population—A real or hypothetical collection of all things of a type. A population may be finite (for example, all wells drilled in 1937) or infinite (all possible bottom-hole tests that have been or might be run).

Probabilistic—Uncertain; containing an element of random chance.

Probability—A means of expressing an outcome on a numerical scale that ranges from impossibility to absolute certainty.

Probability density function—See Density function.

"Proximity" equation—A linear equation, fitted by least squares relating geologic variables measured on well logs to distance from production.

Q—Symbolic representation for the ultimate amount of recoverable petroleum in a region; the petroleum resource base.

Random—Occurring or observed without bias, so the appearance of any value within the range of the variable or process is determined only by chance factors.

Random sample—A subset of a population, selected by a random procedure which insures that any observation has an equal chance of being selected.

Random variable—A variable whose particular values cannot be predicted, but whose behavior is governed by a probability distribution.

Range—The largest minus the smallest of a set of values.

Regional trend—A slowly varying or regional component of a mapped variable, for example as approximated by a polynomial trend surface.

Regionalized variable theory—The theoretical behavior of a variable that does not follow a simple deterministic function, but does possess continuity from point to point and so cannot be regarded as a random variable.

Regression—A statistical method for investigating the relation between a predicted or dependent variable and one or more predictor or independent variables.

Relative frequency—An empirical definition of probability, based on the proportion of times an event is observed in a long series of trials.

Residual—The deviation or difference between a fitted model or equation and the observed value. Widely computed in conjunction with trend surface analysis.

Risk averse—The investment outlook, or utility function, of an individual or organization that places a greater weight on a potential loss than the weight given to a corresponding gain.

Risk indifferent—An investment outlook in which neither potential losses nor gains are given selective weightings with respect to each other.

Risk seeking—An investment outlook in which potential gains are selectively weighted more heavily than corresponding losses.

Root mean squared error—The square root of average squared deviation between a fitted model and the observations. It is equivalent to the variance but not corrected for the mean.

Sample—Observation on part of a population.

Second-degree trend surface—A trend surface generated by an equation having powers up to X^2 and Y^2 and their cross-products.

Semivariance—A measure of spatial autocorrelation used in kriging. Essentially, it is the variance of the differences between points on a surface spaced a distance apart.

Semivariogram—Plot of semivariance versus distance between points being compared.

Significance—A probability measure of the degree of difference between an observed statistic and its value predicted by some model, expressing the likelihood that the observation and prediction are estimates of the same quantity.

Skewness—Asymmetry in a frequency distribution. A unimodal distribution having a tail extending to low values has negative skewness. If the tail extends to high values, the distribution has positive skewness.

Spatial autocorrelation—Measure of the similarity between points on a surface and those a specified distance and direction away. The measure of similarity is the correlation coefficient.

Standard deviation—Measure of the dispersion in measurements of a property about their mean value. It is the square root of the variance.

Stepwise discriminant function—A discriminant function procedure in which every variable is tested for the significance of its contribution to discrimination before being included in the discriminant equation.

Subjective probability—An estimate of the probability of a specific outcome that is an intuitive guess, in contrast to an estimate based on frequencies or on theory.

Target area—An exploratory area in which procedures developed in a "training" area will be applied.

Third-degree trend surface—A trend surface generated by an equation having powers up to X^3 and Y^3 and their cross-products.

Training area—A mature area believed geologically similar to the target area, in which probabilities of exploratory success conditional on geologic conditions can be assessed by reexperiencing the development of the area.

Transformation—A systematic numerical change of a variable, as for example, the conversion to logarithms of its values.

Trend surface analysis—A map analysis procedure in which values of the surface at control points are estimated by a linear equation, usually a polynomial, based on powers and cross-products of the geographic coordinates. Variation in the surface is separated into two components, the trend (regional) and residuals or deviations (local).

Uncertainty function—A function that relates the estimation error in a mapped variable to distance from the nearest control point.

Uncertainty map—See Error map.

Unconditional probability—A probability that does not depend on the outcome of a prior event.

Unit regional value analysis—Calculation of mineral productivity on a unit-of-area basis, as for example, petroleum production per square mile over a region.

Utile—An arbitrary measure of the worth of money to a decision-maker.

Utility function—A curve relating the worth of money to a decision-maker as a function of the amount of money gained or lost.

Variance—Measure of the dispersion in measurements of a property about their mean value. It is the average squared deviation about the mean.

Weighting function—In local fit mapping procedures, the influence of control points is weighted according to their distance away from the point being estimated. A plot of weight versus distance is the weighting function.

References

Abry, C. G., 1973, Quantitative estimation of oil-exploration outcome probabilities in the Tatum Basin, New Mexico: Ph.D. thesis, Stanford University, 151 pp.

Abry, C. G., 1975, Geostatistical model for predicting oil, Tatum Basin, New Mexico: Bulletin of the American Association of Petroleum Geologists, Vol. 59, pp. 2111–2122.

Agterberg, F. P., 1974, Geomathematics: Elsevier Scientific Publishing Co., Amsterdam, 596 pp.

Agterberg, F. P., Chung, C. F., Fabbri, A. G., Kelly, A. M., and Springer, J. S., 1972, Geomathematical evaluation of copper and zinc potential of the Abitibi area, Ontario and Quebec: Geological Survey Canada Paper 71–41, 55 pp.

Ahrens, L. H., 1954, The lognormal distribution of the elements (a fundamental law of geochemistry and its subsidiary): Geochimica et Cosmochimica Acta, Vol. 5, pp. 49–73.

Aitchison, J., and Brown, J. A. C., 1969, The lognormal distribution: With special reference to its uses in economics: University of Cambridge Department of Applied Economics Monographs, No. 5, Cambridge University Press, Cambridge, 176 pp.

Allais, M., 1957, Method of appraising economic prospects of mining exploration over large territories, Algerian Sahara case study: Management Science, Vol. 3, No. 4, pp. 285–347.

Anderson, T. W., 1958, An introduction to multivariate statistical analysis: John Wiley & Sons, Inc., New York, 374 pp.

Arps, J. J., and Roberts, T. G., 1958, Economics of drilling for Cretaceous oil on east flank of Denver-Julesburg Basin: Bulletin of the American Association of Petroleum Geologists, Vol. 42, No. 11, pp. 2549–2566.

Arrow, K. J., 1971, Essays in the theory of risk-bearing: Markham Publishing Co., Chicago, 278 pp.

Batcha, J. P., and Reese, J. R., 1964, Surface determination and automatic contouring for mineral exploration, extraction, and processing: Colorado School Mines Quarterly, Vol. 59, pp. 1–4.

Beebe, B. W., 1959, Characteristics of Mississippian production in the northwestern Anadarko Basin: Tulsa Geological Society Digest, Vol. 27, pp. 190–205.

Benelli, G. C., 1967, Forecasting profitability of oil exploration projects: Bulletin of the American Association of Petroleum Geologists, Vol. 51, pp. 2228–2245.

254 References

Bernoulli, D., 1954, Exposition of a new theory of the measurement of risk: Econometrica, Vol. 22, No. 1, pp. 23–36.

Blackith, R. E., and Reyment, R. A., 1971, Multivariate morphometrics: Academic Press, Inc., London, 412 pp.

Blais, R. A., and Carlier, P. A., 1968, Application of geostatistics in ore evaluation: Canadian Institute of Mining and Metallurgy, Ore Reserve Estimation and Grade Control, Special Vol. 9, pp. 41–68.

Boots, B. N., 1974, Delaunay triangles, an alternative approach to point pattern analysis: Proceedings of the Association of American Geographers, Vol. 6, pp. 26–29.

Breitenbach, E. A., and Peterson, D. W., 1972, Digital log data find oil in wildcat areas: World Oil, Vol. 174, No. 7, June, pp. 71–75.

Brinck, J. W., 1967, Note on the distribution and predictability of mineral resources: European Atomic Energy Community Report EVR 3461.e, 43 pp.

Brock, T. N., 1974, A conceptual Bayesian geostatistical model for metal endowment: M.S. thesis, Department of Mineral Economics, Pennsylvania State University, 133 pp.

Brown, B. W., 1962, Stochastic variables of geologic search and decision: Geological Society of America Bulletin, Vol. 72, pp. 1675–1686.

Celasun, M., 1964a, Some planning problems in mineral exploration: Sc.D. thesis, Columbia University, 202 pp.

Celasun, M., 1964b, The allocation of funds to reconnaissance drilling projects: Colorado School Mines Quarterly, Vol. 59, pp. 169–185.

Chayes, F., 1964, A petrographic distinction between Cenozoic volcanics in and around the open oceans: Journal of Geophysical Research, Vol. 69, No. 8, pp. 1572–1588.

Clair, J. R., 1948, Preliminary notes on lithologic criteria for identification and subdivision of the Mississippian rocks in western Kansas: Kansas Geological Society, Wichita, pp. 1–14.

Cooley, W. W., and Lohnes, P. R., 1971, Multivariate data analysis: John Wiley & Sons, Inc., New York, 364 pp.

Coyle, R. G., 1971, A dynamic model of the copper industry—some preliminary results: Proceedings of the 9th International Symposium on Techniques for Decision-Making in the Mineral Industry, Canadian Institute of Mining & Metallurgy, CIM Special Vol. 12, pp. 53–86.

Crain, I. K., 1970, Computer interpolation and contouring of two dimensional data, a review: Geoexploration, Vol. 8, pp. 71–86.

Dahlberg, E. C., 1975, Relative effectiveness of geologists and computers in mapping potential hydrocarbon exploration targets: Mathematical Geology, Vol. 7, pp. 373–394.

David, M., 1972, Geostatistical ore reserve estimation: 10th International Applications of Computer Methods in the Mineral Industry Symposium, Johannesburg, pp. 27–34.

Davis, J. C., 1973, Statistics and data analysis in geology: John Wiley & Sons, Inc., New York, 550 pp.

Davis, J. C., 1976, Contouring algorithms, in Aangeenbrug, R. T. (Ed.), Proceedings of the 2nd International Symposium on Computer-Assisted Cartography: American Congress on Surveying and Mapping, in press.

Davis, J. C., and Sampson, R. J., 1974, Contouring geologic data by computer (abs.): American Association of Petroleum Geologists Annual Meeting Abstracts, Vol. 1, p. 25.

de Guenin, J., 1962, Exploration and search theory: Computer Short Course and Symposium on Mathematical Techniques and Computer Applications in Mining and Exploration, University of Arizona, College of Mines, Vol. 1, pp. 11–17.

Demirmen, F., 1972, Mathematical procedures and FORTRAN IV program for description of three-dimensional surface configurations: Kansas Geological Survey, Technical Report, KOX Project, University of Kansas, Lawrence, 131 pp.

Demirmen, F., 1973a, Numerical description of folded surfaces depicted by contour maps: Journal of Geology, Vol. 81, No. 5, pp. 599–620.

Demirmen, F., 1973b, Probabilistic study of oil occurrence based on geologic structure in Stafford County, south-central Kansas: Kansas Geological Survey, Technical Report, KOX Project, University of Kansas, Lawrence, 188 pp.

Dowds, J. P., 1961, Mathematical probability as an oil-search tool: World Oil, Vol. 153, No. 4, September, pp. 99–106.

Dowds, J. P., 1964a, Application of information theory in establishing oil field trends: in Parks, G. A. (Ed.), Computers in the Mineral Industries, School of Earth Sciences, Stanford University, Publications in the Geological Sciences, Vol. 9, No. 1, pp. 557–610.

Dowds, J. P., 1964b, Oil finding: A practical problem in statistical decision theory for technologists and management: Colorado School Mines Quarterly, Vol. 59, No. 4, pp. 537–555.

Dowds, J. P., 1965, A decade of successful discovery case histories: World Oil, Vol. 161, No. 4, September, pp. 96–102.

Dowds, J. P., 1966, Petroleum exploration by Bayesian analysis: in Proceedings of the 6th Annual International Symposium on Computer Operations Research, Pennsylvania State University, Vol. 2, pp. FF1–FF26.

Dowds, J. P., 1968, Mathematical probability approach proves successful: World Oil, Vol. 167, No. 7, December, pp. 82–85.

Dowds, J. P., 1969a, Oil rocks, information theory, Markov chains, entropy: 7th International Symposium on Operations Research and Computer Applications in the Mineral Industries, Colorado School Mines Quarterly, Vol. 64, No. 3, pp. 275–293.

Dowds, J. P., 1969b, Statistical geometry of petroleum reservoirs in exploration and exploitation: Journal of Petroleum Technology, Vol. 21, pp. 841–852.

Dowds, J. P., 1972, Geostatistics aid gas, oil discovery and development: World Oil, Vol. 175, No. 1, July, pp. 57–60.

Drew, L. J., 1966, Grid drilling exploration and its application to the search for petroleum: Ph.D. thesis, Pennsylvania State University, 57 pp.

Drew, L. J., 1967, Grid-drilling exploration and its application to the search for petroleum: Economic Geology, Vol. 62, No. 5, pp. 698–710.

Drew, L. J., 1972, Spatial distribution of the probability of occurrence and the value of petroleum: Kansas, an example: Journal of the International Association for Mathematical Geology, Vol. 4, No. 2, pp. 155–171.

Drew, L. J., 1974, Estimation of petroleum exploration success and the effects of resource base exhaustion via a simulation model: U. S. Geological Survey Bulletin 1328, 25 pp.

Drew, L. J., and Griffiths, J. C., 1965, Size, shape and arrangement of some oilfields in the USA: 1965 Symposium, Computer Applications in the Mineral Industries, Pennsylvania State University, pp. FF-1–FF-31.

Ellis, J. R., Harris, D. P., and VanWie, N. H., 1975, A subjective probability appraisal of uranium resources in the state of New Mexico: U. S. Energy Research and Development Administration, Grand Junction Office, unpublished open-file report, 97 pp.

Ellis, R. M., and Blackwell, J. H., 1959, Optimum prospecting plans in mineral exploration: Geophysics, Vol. 24, pp. 344–358.

Emery, J. E., 1954, The application of a discriminant function to a problem in petroleum petrography: Unpublished M.S. thesis, Division of Mineralogy, Pennsylvania State University, 120 pp.

Engel, J. H., 1957, Use of clustering in mineralogical and other surveys: Proceedings of the 1st International Conference on Operations Research, ORSA, Baltimore, pp. 176–192.

Fisher, R. A., 1936, The use of multiple measurements in taxonomic problems: Annals of Eugenics, vii, pp. 179–188.

Galley, J. F., 1958, Oil and geology in the Permian Basin of Texas and New Mexico: in Habitat of Oil, American Association of Petroleum Geologists, Tulsa, pp. 395–446.

Gotautas, V. A., 1963, Quantitative analysis of prospect to determine whether it is drillable: Bulletin of the American Association of Petroleum Geologists, Vol. 47, No. 10, pp. 1794–1812.

Graybill, F. A., 1961, An introduction to linear statistical models, Vol. 1: McGraw-Hill Book Co., New York, 463 pp.

Grayson, C. J., Jr., 1960, Decisions under uncertainty: Drilling decisions by oil and gas operators: Harvard Business School, Division of Research, Cambridge, Mass., 402 pp.

Grender, G. C., Rapoport, L. A., and Segers, R. G., 1974, Experiment in quantitative geologic modeling: Bulletin of the American Association of Petroleum Geologists, Vol. 58, No. 3, pp. 488–498.

Greysukh, V. L., 1966, The possibility of studying landforms by means of digital computer: Izvestiya Akadimii Nauk SSSR, seriya geograficheskaya, No. 4, pp. 102–110.

Griffiths, J. C., 1962, Frequency distributions of some natural resource materials: 23rd Technical Conference on Petroleum Production, Pennsylvania State University College of Earth and Mineral Science Experiment Station Circular, No. 63, pp. 174–198.

Griffiths, J. C., 1966, Exploration for natural resources: Operations Research, Vol. 14, No. 2, pp. 189–209.

Griffiths, J. C., 1969, The unit regional-value concept and its application to Kansas: Kansas Geological Survey Special Distribution Publication 38, 48 pp.

Griffiths, J. C., and Drew, L. J., 1964, Simulation of exploration programs for natural resources by models: Colorado School Mines Quarterly, Vol. 59, No. 4, pp. 187–206.

Griffiths, J. C., and Drew, L. J., 1966, Grid spacing and success ratios in exploration for natural resources: 6th Annual Symposium and Short Course on Computers and Operations Research in Mineral Industries, Pennsylvania State University, 17 pp.

Griffiths, J. C., and Singer, D. A., 1971, A first generation simulation model for selecting among exploration programs, with special application to the search for uranium ore bodies: Geocom Programs, No. 2, London, 42 pp.

Griffiths, J. C., and Singer, D. A., 1972, The Engel simulator and the search for uranium: in Salamon, M. D. G., and Lancaster, F. H. (Eds.), Application of Computer Methods in the Mineral Industry: Proceedings of the 10th International Symposium, South African Institute of Mining and Metallurgy, pp. 9–16.

Hambleton, W. W., Davis, J. C., and Doveton, J. H., 1975, Estimating exploration potential, in Haun, J. D. (Ed.), Methods of Estimating the Volume of Undiscovered Oil and

Gas Resources: American Association of Petroleum Geologists, Studies in Geology No. 1, pp. 171–185.

Harbaugh, J. W., and Bonham-Carter, G., 1970, Computer simulation in geology: Wiley-Interscience, New York, 575 pp.

Harbaugh, J. W., and Merriam, D. F., 1968, Computer applications in stratigraphic analysis: John Wiley & Sons, Inc., New York, 282 pp.

Harder, R. L., and Desmarais, R. N., 1972, Interpolation using surface splines: Journal of Aircraft, Vol. 9, pp. 189–191.

Harris, D. P., 1964, The occurrence of mineral wealth: American Institute of Mining, Metallurgical, and Petroleum Engineers, 1964 Annual Meeting (preprint), 13 pp.

Harris, D. P., 1965a, An application of multivariate statistical analysis to mineral exploration: Ph.D. thesis, Pennsylvania State University, 261 pp.

Harris, D. P., 1965b, Multivariate statistical analysis—a decision tool for mineral exploration: Short Course and Symposium on Computers and Computer Applications in Mining and Exploration, Vol. 1: University of Arizona, pp. C1–C35.

Harris, D. P., 1968, Potential mineral reserves of Seward Peninsula, Alaska: An evaluation by geostatistics and computer simulation: Mineral Industry Research Laboratory, MIRL Bull. 18, Part II, University of Alaska, College, Alaska, pp. 60–105.

Harris, D. P., 1969a, Alaska's base and precious metals resources: A probabilistic regional appraisal: 7th International Symposium on Operations Research and Computer Applications in the Mineral Industries, Colorado School Mines Quarterly, Vol. 64, No. 3, pp. 295–327.

Harris, D. P., 1969b, Quantitative methods, computers, reconnaissance geology and economics in the appraisal of mineral potential: International Computer Applications Symposium, Salt Lake City, Utah, September 17–19, 43 pp.

Harris, D. P., and Chrow, J. K., 1969, Alaska's base and precious metals resources: A probabilistic regional appraisal: Part I, Geology, production, and method: Department of Mineral Economics, Pennsylvania State University, unpublished report, 34 pp.

Harris, D. P., Freyman, A. J., and Barry, G. S., 1971, A mineral resource appraisal of the Canadian Northwest using subjective probabilities and geological opinion: Proceedings of the 9th International Symposium on Techniques for Decision-Making in the Mineral Industry, Canadian Institute of Mining and Metallurgy, CIM Special Vol. No. 12, pp. 100–116.

Haun, J. D., 1971, Potential oil and gas resources: Colorado School Mines, unpublished progress report, 18 pp.

Hewitt, C. H., 1966, How geology can help engineer your reservoirs: Oil and Gas Journal, Vol. 64, No. 46, November, pp. 171–178.

Hubbert, M. K., 1967, Degree of advancement of petroleum exploration in the United States: Bulletin of the American Association of Petroleum Geologists, Vol. 51, pp. 2207–2227.

Huijbregts, C. J., 1975, Regionalized variables and quantitative analysis of spatial data: *in* Davis, J. C. and McCullagh, M. J. (Eds.), Display and Analysis of Spatial Data, John Wiley & Sons, Ltd., London, pp. 38–53.

Huijbregts, C. J., and Journel, A., 1972, Estimation of lateritic-type ore bodies: 10th International Symposium on Applications of Computer Methods in the Mineral Industry, Johannesburg, pp. 207–212.

Huijbregts, C. J., and Matheron, G., 1971, Universal Kriging: an optimal method for contouring and trend surface analysis: Proceedings of the 9th International Symposium on Techniques for Decision-Making in the Mineral Industry, Canadian Institute of Mining and Metallurgy, CIM Special Vol. No. 12, pp. 159–169.

Huijbregts, C. J., and Segovia, R., 1973, Geostatistics for the valuation of a copper deposit: 11th International Symposium on Applications of Computer Methods in the Mineral Industry, Tucson, Arizona, pp. D24–D43.

IBM, 1965, Numerical surface techniques and contour map plotting: IBM Data Processing Application, White Plains, New York, 36 pp.

Imai, S., and Itho, S., 1971, Some techniques for the determination of effective drill spacing: Proceedings of the 9th International Symposium on Techniques for Decision-Making in the Mineral Industry, Canadian Institute of Mining and Metallurgy, CIM Special Vol. No. 12, pp. 199–208.

Irwin, C. D., 1971, Stratigraphic analysis of Upper Permian and Lower Triassic strata in southern Utah: Bulletin of the American Association of Petroleum Geologists, Vol. 55, pp. 1976–2007.

Jones, R. L., 1971, A generalized digital contouring program: NASA Langley Research Center, Hampton, Virginia, NASA TN D-6022, 78 pp.

Kaufman, G. M., 1963, Statistical decision and related techniques in oil and gas exploration: Prentice-Hall, Englewood Cliffs, 307 pp.

Kaufman, G. M., 1974, Statistical methods for predicting the number and size distribution of undiscovered hydrocarbon deposits: American Association of Petroleum Geologists Research Symposium on Methods of Estimating Volume of Undiscovered Oil and Gas Resources, Stanford University, August 21–23, pp. 247–310.

Kaufman, G. M., Balcer, Y., and Kruyt, D., 1975, A probabilistic model of oil and gas discovery: in Haun, J. D. (Ed.), Methods of Estimating the Volume of Undiscovered Oil and Gas Resources, American Association of Petroleum Geologists, Studies in Geology No. 1, pp. 113–142.

Kendall, M. G., 1968, A course in multivariate analysis: Hafner Publishing Co., New York, 185 pp.

Kendall, M. G., and Moran, P. A. P., 1963, Geometrical probability: Griffin's statistical monographs and courses, No. 10: Hafner Publishing Co., New York, 125 pp.

Kendall, M. G., and Stuart, A., 1967, The advanced theory of statistics: Vol. 3, Design and analysis, and time series: Charles Griffin and Company Ltd., London, 557 pp.

Kendall, M. G., and Stuart, A., 1969, The advanced theory of statistics: Vol. 1, Distribution theory (3rd edition): Charles Griffin and Company Ltd., London, 439 pp.

Koch, G. S., Jr., and Link, R. F., 1970, Statistical analysis of geological data: John Wiley & Sons, Inc., New York, 375 pp.

Koopman, B., 1956a, The theory of search, Part 1, Kinematic bases: Operations Research, Vol. 4, pp. 324–346.

Koopman, B., 1956b, The theory of search, Part 2, Target detection: Operations Research, Vol. 4, pp. 503–531.

Koopman, B., 1957, The theory of search, Part 3, Optimum distribution of searching effort: Operations Research, Vol. 5, pp. 613–626.

Krumbein, W. C., and Graybill, E. A., 1965, An introduction to statistical models in geology: McGraw-Hill Book Co., New York, 475 pp.

Krumbein, W. C., and Pettijohn, F. J., 1938, Manual of sedimentary petrography: Appleton-Century-Crofts, Inc., New York, 549 pp.

Kyrala, A., 1964, Basic probabilistic methods in geological search: Geophysics, Vol. 29, pp. 105–108.

Lampietti, F. J., and Marcus, L. F., 1971, Computer simulation of Ringarooma Bay offshore sampling for tin: Proceedings of the 9th International Symposium on Techniques for Decision-Making in the Mineral Industry, Canadian Institute of Mining and Metallurgy, CIM Special Vol. No. 12, pp. 223–228.

Marriott, F. H. C., 1974, The interpretation of multiple observations: Academic Press, London, 117 pp.

Marshall, K. T., 1964, A preliminary model for determining the optimum drilling pattern in locating and evaluating an ore body: Colorado School Mines Quarterly, Vol. 59, pp. 223–236.

Martin, J. J., 1967, Bayesian decision problems and Markov chains: John Wiley & Sons, Inc., New York, 202 pp.

Mast, R. F., 1970, Size, development, and properties of Illinois oilfields: Illinois Geological Survey, Illinois Petroleum 93, 42 pp.

Matheron, G., 1965, Les variables régionalisées et leur estimation: Une application de la théorie des fonctions aléatoires aux sciences de la nature: Masson et Cie, Editeurs, Paris, 305 pp.

Matheron, G., 1969, Le Krigeage universel: Les Cahiers du Centre de Morphologie Mathématique de Fontainebleau, Vol. 1, 83 pp.

Matheron, G., 1971, The theory of regionalized variables and its applications: Les Cahiers du Centre de Morphologie Mathématique de Fontainebleau, Vol. 5, 211 pp.

McCrossan, R. G., 1969, An analysis of size frequency distribution of oil and gas reserves of western Canada: Canadian Journal of Earth Science, Vol. 6, No. 2, pp. 201–211.

Megill, R. E., 1971, An introduction to exploration economics: Petroleum Publishing Co., Tulsa, 159 pp.

Menard, H. W., and Sharman, G., 1975, Scientific uses of random drilling models: Science, Vol. 190, pp. 337–343.

Merriam, D. F., 1963, The geologic history of Kansas: Kansas Geological Survey Bulletin 162, 317 pp.

Mickey, M. R., and Jesperson, H. W., 1954, Some statistical problems of uranium exploration: U. S. Atomic Energy Commission, Report RME-3105, 78 pp.

Middleton, G. V., 1962, A multivariate statistical technique applied to the study of sandstone composition: Transactions of the Royal Society of Canada, Vol. 26, Series III, Sec. III, pp. 119–226.

Miller, B. M., and others, 1975, Geological estimation of undiscovered recoverable oil and gas resources in the United States: U. S. Geological Survey Circular 725, 78 pp.

Morgan, B. W., 1968, An introduction to Bayesian statistical decision processes: Prentice-Hall, Inc., Englewood Cliffs, 116 pp.

Morrison, D. F., 1967, Multivariate statistical methods: McGraw-Hill Book Co., New York, 338 pp.

Newendorp, P. D., 1967, Application of utility theory to drilling investment decisions: D. Engr. thesis, University of Oklahoma, 145 pp.

Newendorp, P. D., 1968, A decision criterion for drilling investments: Society of Petroleum Engineers Preprint SPE 2219, SPE 43rd Annual Fall Meeting, Houston, Texas.

Newendorp, P. D., 1976, Decision analysis for petroleum exploration: Petroleum Publishing Co., Tulsa, 750 pp.

Nordbeck, S., and Rystedt, B., 1972, Computer cartography: Studentlitteratur Lund, Lund, Sweden, 315 pp.

Olea, R. A., 1974, Optimal contour mapping using universal kriging: Journal of Geophysical Research, Vol. 79, No. 5, pp. 695–702.

Olea, R. A., 1975, Optimum mapping techniques using regionalized variable theory: Kansas Geological Survey, Series on Spatial Analysis No. 2, 137 pp.

Olea, R. A., and Davis, J. C., 1977, Use of regionalized variables for evaluation of petroleum occurrence in the Magellan Basin of South America: Bulletin of the American Association of Petroleum Geologists, in press.

Osborn, R. T., 1967, An automated procedure for producing contour charts: U. S. Naval Oceanographic Office, IM No. 67-4, 54 pp.

Overall, J. E., and Klett, C. J., 1972, Applied multivariate analysis: McGraw-Hill Book Co., New York, 500 pp.

Pachman, J. M., 1966, Optimization of seismic reconnaissance surveys in petroleum exploration: Management Science, Vol. 12, No. 8, pp. B312–322.

Palmer, J. A. B., 1969, Automated mapping: Proceedings of the 4th Australian Computer Conference, Adelaide, South Australia, pp. 463–466.

Pratt, J. W., 1964, Risk aversion in the small and in the large: Econometrica, Vol. 32, pp. 122–136.

Prelat, A., 1974, Statistical estimation of wildcat well outcome probabilities by visual analysis of structure contour maps of Stafford County, Kansas: Ph.D. thesis, Stanford University, 103 pp.

Raiffa, H., 1970, Decision analysis: Introductory lectures on choices under uncertainty: Addison-Wesley Publishing Co., Reading, Mass., 309 pp.

Rhynsburger, D., 1973, Analytic delineation of Thiessen polygons: Geographical Analysis, Vol. 5, pp. 134–144.

Robinson, J. E., 1975, Frequency analysis, sampling, and errors in spatial data: in Davis, J. C. and McCullagh, M. J. (Eds.), Display and Analysis of Spatial Data: John Wiley & Sons, Ltd., London, pp. 78–95.

Robinson, J. E., Charlesworth, H. A. K., and Ellis, M. J., 1969, Structural analysis using spatial filtering in Interior Plains of south-central Alberta: Bulletin of the American Association of Petroleum Geologists, Vol. 53, No. 11, pp. 2341–2367.

Robinson, J. E., and Ellis, M. J., 1971, Spatial filters and FORTRAN IV program for filtering geologic maps: Geocom Programs, No. 1, London, 21 pp.

Salisbury, G. P., 1968, Natural gas in Devonian and Silurian rocks of the Permian Basin, West Texas and southeast New Mexico: American Association of Petroleum Geologists, Memoir 9, Vol. 2, pp. 1433–1445.

Sampson, R. J., 1975a, The SURFACE II graphics system: in Davis, J. C. and McCullagh, M. J. (Eds.), Display and Analysis of Spatial Data: John Wiley & Sons, Ltd., London, pp. 244–266.

Sampson, R. J., 1975b, SURFACE II graphics system: Kansas Geological Survey, Series on Spatial Analysis No. 1, 240 pp.

Savinskii, I. D., 1965, Probability tables for locating elliptical underground masses with a rectangular grid: Consultants Bureau, Plenum Press, New York (trans. from Russian), 100 pp.

Schlaifer, R., 1961, Introduction to statistics for business decisions: McGraw-Hill Book Co., New York, 382 pp.

Schwade, I. T., 1967, Geologic quantification: description numbers success ratio: Bulletin of the American Association of Petroleum Geologists, Vol. 51, No. 7, pp. 1225–1239.

Sinclair, A. J., and Woodsworth, G. J., 1970, Multiple regression as a method of estimating exploration potential in an area near Terrace, British Columbia: Economic Geology, Vol. 65, pp. 998–1003.

Singer, D. A., 1971, Multivariate statistical analysis of the unit regional value of mineral resources: Ph.D. thesis, Department of Geochemistry and Mineralogy, Pennsylvania State University, 210 pp.

Singer, D. A., 1972, Elipgrid, a FORTRAN IV program for calculating the probability of success in locating elliptical targets with square, rectangular and hexagonal grids: Geocom Programs, No. 4, London, 16 pp.

Singer, D. A., 1975, Relative efficiencies of square and triangular grids in the search for elliptically shaped resource targets: U. S. Geological Survey, Journal of Research, Vol. 3, No. 2, pp. 163–167.

Singer, D. A., and Wickman, F. E., 1969, Probability tables for locating elliptical targets with square, rectangular and hexagonal point-nets: Pennsylvania State University Mineral Science Experiment Station Special Publication 1–69, 100 pp.

Slichter, L. B., Dixon, W. J., and Myer, G. H., 1962, Statistics as a guide to prospecting: Computer Short Course and Symposium on Mathematical Techniques and Computer Applications in Mining and Exploration, University of Arizona, College of Mines, Vol. 1, pp. F-1–F-27.

Smith, M. B., 1972, Parametric utility functions for decisions under uncertainty: Society of Petroleum Engineers Preprint SPE 3973, SPE 47th Annual Fall Meeting, San Antonio, Texas, 11 pp.

Stachtchenko, L., 1971, Aerial survey strategies for copper exploration: Proceedings of the 9th International Symposium on Techniques for Decision-Making in the Mineral Industry, Canadian Institute of Mining and Metallurgy, CIM Special Vol. No. 12, pp. 117–129.

Uhler, R. S., and Bradley, P. G., 1970, A stochastic model for determining the economic prospects of petroleum exploration over large regions: Journal of the American Statistical Association, Vol. 65, No. 330, Applications Section, pp. 623–630.

Von Neumann, J., and Morgenstern, O., 1947, Theory of games and economic behavior: Princeton University Press, Princeton, New Jersey, 641 pp.

Wagner, H., 1969, Principles of operations research with applications to managerial decisions: Prentice-Hall, Inc., Englewood Cliffs, New Jersey, 937 pp.

Walters, R. F., 1969, Contouring by machine: A user's guide: Bulletin of the American Association of Petroleum Geologists, Vol. 35, No. 11, pp. 2324–2340.

White, D. A., Garrett, R. W., Jr., Marsh, G. R., Baker, R. A., and Gehman, H. M., 1975, Assessing regional oil and gas potential: in Haun, J. D. (Ed.), Methods of Estimating the Volume of Undiscovered Oil and Gas Resources, American Association of Petroleum Geologists, Studies in Geology No. 1, pp. 143–159.

Wren, A. E., 1975, Contouring and the contour map, a new perspective: Geophysical Prospecting, Vol. 23, pp. 1–17.

Index

263